SpringerBriefs in Molecular Science

History of Chemistry

Series editor

Seth C. Rasmussen, Fargo, North Dakota, USA

More information about this series at http://www.springer.com/series/10127

Emma Tobin · Chiara Ambrosio
Editors

Theory Choice in the History of Chemical Practices

 Springer

Editors
Emma Tobin
Department of Science and Technology
 Studies
University College London
London
UK

Chiara Ambrosio
Department of Science and Technology
 Studies
University College London
London
UK

ISSN 2191-5407   ISSN 2191-5415   (electronic)
SpringerBriefs in Molecular Science
ISSN 2212-991X
SpringerBriefs in History of Chemistry
ISBN 978-3-319-29891-7   ISBN 978-3-319-29893-1   (eBook)
DOI 10.1007/978-3-319-29893-1

Library of Congress Control Number: 2016934673

Printed on acid-free paper

This Springer imprint is published by Springer Nature
The registered company is Springer International Publishing AG Switzerland

Contents

Editors and Contributors

About the Editors

Emma Tobin is a lecturer in Philosophy of Science at the Department of Science and Technology Studies at University College London. She concluded her Ph.D. in Philosophy at Trinity College Dublin in 2005. She has published many articles in Philosophy of Science, particularly in relation to the topic of classification in science. Her research interests include history and philosophy of science, metaphysics of science and philosophy of chemistry.

Chiara Ambrosio is a lecturer in History and Philosophy of Science at University College London. Her research focuses on the relations between art and science in the nineteenth and twentieth centuries, American pragmatism and the philosophy of Charles Sanders Peirce. She ties these research interests to broader themes in history and philosophy of science, particularly to debates around scientific representations and the visual culture(s) of science.

About the Contributors

Kat F. Austen is a succession of experiences and an assemblage of aspirations. She is also a person. In the temporal melting-pot of her life so far she has produced work as an artist, an environmental scientist, a writer and much in between. She explores networks of unseen influence and aims to understand truth-seeking and collective actions. Kat is Artist in Residence at the Faculty of Maths and Physical Sciences, University College London, Head of Research and Design at social enterprise iilab. She holds a Ph.D. in Computational Chemistry from University College London and worked as a post-doctoral research associate at the University of Cambridge. Her writing has appeared in New Scientist, Nature, The Ecologist and The Guardian, and she consults widely on the intersection of science, society, art and technology, including as a Future shaper for Forum for the Future, for the European Commission and Ofwat.

Georgiana D. Hedesan is a Wellcome Trust Fellow in Medical Humanities at the University of Oxford, researching the topic of universal medicine in seventeenth-century alchemy. Her first monograph, entitled An Alchemical Quest for Universal Knowledge: The 'Christian Philosophy' of Jan Baptist Van Helmont (1579–1644) will be published in June 2016 with Ashgate Press. Prior to her employment at Oxford, Georgiana (Jo) was a short-term Frances A. Yates Fellow at the Warburg Institute, London, and a Cantemir Junior Fellow at the University of Oxford. She concluded her Ph.D. in History at University of Exeter at the end of 2012 with a dissertation on Jan Baptist Van Helmont. She has published articles in such scholarly journals as Medical History and Ambix, as well as chapter contributions in edited books.

Matthew Paskins recently completed his Ph.D. in University College London's Department for Science and Technology Studies. His main research interests include connections between useful knowledge and material properties in Britain during the long eighteenth century: which means grass seeds, tree accounts, potash, iron sand and varnish trees. He is currently Community Manager for the Open University/Arts and Humanities Research Council "Commodity Histories" project and Consultant Historian for the Tree Council's "Hedgerow Harvest" oral history project.

Jennifer M. Rampling is Assistant Professor of History at Princeton University, where she teaches Program in History of Science. She holds a Ph.D. in History and Philosophy of Science from Cambridge University. A specialist in the history of late medieval and early modern alchemy, she is currently completing her first book on the history of English alchemy. She is the Editor of Ambix, the main journal for the history of alchemy and chemistry.

Georgette Taylor earned a Ph.D. in 2006 on the chemical affinity theories of the eighteenth century and explored the teaching of chemistry, in particular by William Cullen and Joseph Black and by many of their ex-students. She won the 2008 Partington Prize awarded by the Society for the History of Alchemy and Chemistry for 'Tracing Influence in Small Steps: Richard Kirwan's Quantified Affinity Theory'. A post-doctoral fellowship followed, with the project 'Analysis and Synthesis in 19th Century Chemistry: Towards a New Philosophical History of Scientific Practice'.

Contributors

Chiara Ambrosio Department of Science and Technology Studies, University College London, London, UK

Kat F. Austen Artist in Residence, Faculty of Mathematical and Physical Sciences, University College London, London, UK

Georgiana D. Hedesan Wolfson College, University of Oxford, Oxford, UK

Matthew Paskins "Commodity Histories" Project, Open University, Milton Keynes, UK

Jennifer M. Rampling Department of History, Princeton University, Princeton, NJ, USA

Georgette Taylor Department of Science and Technology Studies, University College London, London, UK

Emma Tobin Department of Science and Technology Studies, University College London, London, UK

Abstract

This collection of articles aims to study theory choice in the context of some key theoretical developments in the history of chemistry. The analysis shows that *theories*—however defined from a philosophical point of view—were first and foremost used in the context of particular, historically contingent practices pursued by specific communities and historical actors. The main goal of this volume is to bring together a *history of chemical practices*, and in so doing reveal that theory choice is conceptually more problematic than was originally conceived.

Keywords History of chemistry · Chemical practices · Theory choice

Chapter 1
Introduction

Emma Tobin and Chiara Ambrosio

This collection of articles aims to study theory choice in the context of some key theoretical developments in the history of chemistry. Perhaps, the most influential account of theory choice in the philosophy of science has been Thomas Kuhn's (1970) account in *The Structure of Scientific Revolutions* [1]. Kuhn himself contributed to reframe the philosophical problem of theory choice in light of the insights we may derive from thorough historical investigation. Just over fifty years since the publication of Kuhn's seminal book, this volume proposes to engage with his philosophical agenda with a renewed historical and historiographical awareness.

Theory choice for Kuhn was dictated by the problem of anomalies in the context of a background scientific paradigm. An anomaly, which would remain ever recalcitrant to explanation within the resources of a background theory, would eventually bring about a crisis, leading to the articulation of a new theory to explain the anomaly. Kuhn used the history of science and indeed the history of chemistry, to motivate this account of theory change. This account makes theory change absolute in so far as the choice to adopt the new theory entails the rejection of the old theory; namely in Kuhnian terms theories are incommensurable across periods of scientific change, which he defined as "scientific revolutions". Kuhn used the 18th century chemical revolution as one of his archetypal historical examples of theory change. Moreover, on the Kuhnian picture, the theory-dependence of all observation entails that there is no theory-neutral way of observing in chemical practice (and in science more broadly). Theory choice is necessitated by the fact that competing theories are mutually exclusive.

E. Tobin (✉) · C. Ambrosio
Department of Science and Technology Studies, University College London,
Gower Street, London WC1E 6BT, UK
e-mail: e.tobin@ucl.ac.uk

C. Ambrosio
e-mail: c.ambrosio@ucl.ac.uk

© Springer International Publishing Switzerland 2016
E. Tobin and C. Ambrosio (eds.), *Theory Choice in the History
of Chemical Practices*, SpringerBriefs in History of Chemistry,
DOI 10.1007/978-3-319-29893-1_1

This volume aims to revisit that question,[1] but it does so by looking at the history of chemistry through the lens of a number of different periods, towards a more comprehensive historical analysis of the question of theory in the history of chemical practice.[2] The consensus picture that emerges is that the history of science tells a much more complex story about theory choice. At times, there appears to be a number of operative theories that are not as such competing, but the choice of which theory to adopt is dictated by the context of scientific practice at the time. In these cases, there appears to be a tolerance of competing theories, at one and the same time, which presents a challenge to the Kuhnian picture. At other times, the direct opposite appears to be true: some theories or available alternative explanations are simply not selected at all in the historical narrative. Thus there seem to be, at least at certain points in history, particular selective biases in favour of theories which gain consensus and turn out to dominate for social, as well as epistemological, reasons.

A glimpse at history shows that *theories*—however defined from a philosophical point of view—were first and foremost used in the context of particular, historically contingent practices pursued by specific communities and historical actors. For instance, they sometimes played a pivotal pedagogical role in training the next generation of chemists (e.g. affinity theory—Chap. 4). Thus, with its focus on chemical *practice*, this volume challenges many of the standard ways in which philosophers approached the issue of theory choice. Neither revolutions (and with them discontinuity across paradigms), nor continuity across theory change (as portrayed in some realist accounts of theory choice) seem to fully accommodate the variety of motivations and contexts that our cases aim to bring to the attention of philosophers, historians and scientists alike. Indeed, the underlying motive of our account of theory choice is that any attempt at resolving the conceptual challenges arising from it needs to be formulated against the background of history, and against the needs, goals and aims that animated scientific practice at particular times in history. It is in this respect that the main goal of this volume is to bring together a *history of chemical practices*, and in so doing revealing that theory choice is conceptually more problematic than was originally conceived.

In Chap. 2, Jennifer Rampling begins the discussion with the context of medieval alchemy. Judged from our present standpoint, alchemy is often presented as pre-scientific, since alchemists aspired to "impossible" transformations. Moreover, the methods used in the attempts to perform these transformations were numerous

---

[1]The papers in this volume were presented over the course of 2012/2013 at the AdHoc London meetings by members of the AdHoc London and Cambridge group, jointly organized by the Science and Technology Studies Department in University College London and the Department of History and Philosophy of Science in Cambridge.

[2]It is partly for this reason that we have decided not to focus on the much-debated case of the Chemical Revolution, covered by Kuhn. Our aim is to move to less investigated areas in the history of chemistry, and explore the insights that those histories might have to offer to the debate. The most up-to date account of the chemical revolution that takes into account historical, historiographical and philosophical issues, including debates related to theory choice is Chang [2].

and heterogeneous in nature, and so it might be claimed that alchemy is "pre-theoretical". Looking at the status of alchemy from a historical standpoint, however, reveals a much more interesting, and epistemically fruitful picture: it is precisely in the context of the competing explanations characterising medieval alchemy, Rampling argues, that the issue of theory choice becomes conceptually challenging and worthy of further investigation.

Rampling questions the degree to which practice is guided by theory in medieval alchemy. She argues that theories are not objectively formulated as a set of hypotheses, but rather they play a mediating role between authority and practice. Thus, the "choice" of theory is dictated by the records of previously successful predecessors in terms of the choice of ingredients and processes used in scientific practice. There is a feedback loop between the two aspects because authority is capable of interpretation in the light of practice. Thus, scientific practice reveals a dynamic process of theoretical refinement in the context of experimentation, even though the initial hypothesis to be tested and the context of testing (e.g. material used etc.) are highly conservative and dictated by authority.

In Chap. 3, Jo Hedesan addresses Boyle's *Sceptical Chymist* as a case study for theory choice in the 17th century. The original discussion focuses on three competing theories concerning the chemical components of matter: the four-element theory of Aristotle, the three-principle theory of Paracelsus and the atomistic theory. Hedesan argues, that in effect the book is an attack on Paracelsianism by Boyle. This is evident from the fact that the other two "theories" fall out of the discussion quite early in the text. In the history of Chemistry, this seminal text is often taken as a successful attack on Paracelsianism.

Hedesan examines some of the arguments canvassed against Paracelsianism in Chap. 3 and she concludes that none of the arguments were sufficient to reject the theory. Moreover, there was no "crucial experiment" used by Boyle to support his rejection either. Rather, Hedesan argues that Boyle was committed to an alternative paradigmatic background, namely Helmontianism. Hedesan suggests that the problem of theory choice is often not rationally grounded, but rather the "choice" of theory and the rejection of other theories can be motivated by the influence of historical actors' commitment to a background paradigm or school of thought.

Chapter 4 features 18th century affinity theory as a way of directly problematizing theory choice in the Kuhnian sense. Georgette Taylor questions the common-sense view that affinity theory was merely a transitional phase between two scientific revolutions. Historians have either emphasized the Newtonian origins of affinity theory, in a sort of assimilation of later events to the sense of authority and validation deriving from the Newtonian paradigm, or have construed affinity theory as a sort of anticipation—often based on a narrow view of the role of phlogiston chemistry—and the great changes brought about by the Chemical Revolution.

What Taylor's account reveals, in contrast to mainstream historical interpretations, is that affinity theory constituted the fundamental core of 18th century chemistry, thus standing as a definitive theory in its own right. Taylor relates this claim to the widespread presence of affinity tables and their indispensable role as pedagogical tools and as guidelines for chemical practice. From being largely ignored when first introduced by Geoffroy in 1719, affinity tables became ubiquitous in 18th century chemistry and showed remarkable resilience to the revolutionary changes in nomenclature and discoveries of new airs that characterized the 19th century.

There is something peculiar about the nature of affinity tables that will puzzle the historian, the philosopher, and the chemist alike. On the one hand, they seem to undermine the Kuhnian view that pedagogy at best provides a way of articulating the paradigm further, without providing opportunities for revolutionary change. The articulation of affinity theory reveals a far more complex picture, in which affinity tables provided the necessary continuity to bridge the Kuhnian gap between subsequent (and mutually incompatible) paradigms. Taylor shows this clearly in her discussion of how the doctrine of affinity survived the Chemical Revolution pretty much unscathed.

On the other hand, the generative power of affinity tables is evident in the fact that they were constantly amended and refined, often in light of empirical, tacit and extra-theoretical assumptions. The flexibility of affinity theory, evidenced by the variations upon affinity tables and their diverse uses, was the key to its success. More importantly, this supports the role of affinity theory in the disciplinary development of chemistry as a whole. Affinity had both explanatory and heuristic power, and its pedagogical applications through affinity tables allowed it to be used as an undisputed instrument of the greatest utility, as well as an object of scientific inquiry in its own right.

In Chap. 5, theory choice is explored through the lens of chemical substitution. Drawing on Maxine Berg's formulation of "imitative invention", Matthew Paskins argues that the idea of substitution discloses important avenues of inquiry into how 18th century chemists made sense of the properties of materials. Along with showing the continuity between chemistry, natural history and natural philosophy, substitution features in Paskins' account as a most powerful drive for innovation and change in chemical practice.

Paskins engages with a particular strand in the scholarship on 18th century chemistry that regards knowledge of materials as the primary drive in chemical taxonomy and classification. In response to this literature, and drawing on a version of 'bundle theory' derived from studies in commodity history, Paskins' account places the practice of substitution at the center of chemical knowledge instead. Looking at substitution reveals that 18th century chemists approached materials as 'bundles of properties' and as multi-dimensional objects of inquiry. This explicitly clashes with the common-sense view that taxonomic practices—in chemistry as well as in other branches of science—are an inherently orderly affair. Through a comparison with natural history, Paskins shows that in both fields nomenclature and classification were disorderly in nature, and that this is exactly why philosophers, historians and scientists should pay attention to their role in the formation and consolidation of scientific knowledge.

The discussion here discloses an important angle on a neglected aspect of theory choice in 18th century chemistry. Here we are no longer facing the question of two empirically equivalent theories between which a definite choice needs to be made, nor are we grappling with the question of what rescues the rationality of theory choice from the dangers of a radical account of incommensurability. On the contrary, an analysis of the practices of substitution and classification shows that often theory choice in 18th century chemistry consisted of a quest to locate properties— desirable *and* undesirable ones—that would eventually reveal either new opportunities or usefully point toward dead ends in chemical practice.

In Chap. 6, the issue of theory choice is brought to the fore in the light of the modern practice of computational modeling. In this chapter Kat Austen casts light on another important debate in recent philosophy of science: what is the relationship between theories and the models scientists use to construct, manipulate and represent them? The case of the development of chemistry from the use and incorporation of computational models focuses on the relationship between theory and model building in chemical practice.

The question of "choice" comes into play in many places in the practice of constructing computational models. Once a chemical problem has been identified, there are a number of choices made by the practitioner: a model is chosen, a particular computer code is decided upon and the practitioner decides between the variables that can be fine-tuned within the model (e.g. basic sets, interatomic potentials or optimization algorithm). Austen rightly points out that there is a meta-level choice before we even begin the models-based approach, and that is whether computational modeling should be used at all given the nature of the chemical problem.

This 'choice' is also illustrated in the context of the history of the reception and appropriation of computational models in Chemistry since the 1950s. Some chemists have questioned the reliability of computational methods, either because they are based on simulations, which aim to measure systems that cannot be easily captured experimentally, or because to the extent that they do capture empirical systems, they involve abstractions and so lack the complexity of real life systems. Perhaps, even more worrying is the fact that the scale of chemical interrogation of these systems requires electronic and quantum effects which increase the number of approximations necessary.

The papers in this volume interrogate the complex way in which chemical practice has guided, and still guides, theory choice in chemical practice. At times, it appears that competing theories can coexist, with 'choices' being motivated by the context of use. This is perhaps the key message our contributions convey: theory choice rarely occurs in the artificial and over-simplistic scenario of two empirically equivalent theories. It is instead the context of the *use* of the theory in a community of practitioners which dictates its applicability and indeed perhaps even its success at particular times. Alchemical practice located "choice" at the intersection between practice and authority. The practices of substitution in the 18th century were guided by the quest to locate properties—desirable *or* not, which would allow new chemical opportunities or conversely, to reject theoretical claims because of their practical implications. Equally, affinity theories were chosen not because of their truth or greater truth in comparison with empirically equivalent alternatives, but because of their pedagogical and heuristic value in chemical practice. All of the papers in this collection reveal that this is not merely an abstract historical or philosophical issue per se, but that a historical understanding of chemical practice provides new ways of conceptualising the philosophical problem of theory choice.

References

1. Kuhn T (1970) The structure of scientific revolutions, 2nd edn. University of Chicago Press, Chicago
2. Chang H (2012) Is water H_2O? Evidence, pluralism and realism. Boston studies in the philosophy of science, Springer, Dordrecht, Chapter 1

Chapter 2
Theory Choice in Medieval Alchemy

Jennifer M. Rampling

Theory choice is not a term often used in the context of medieval alchemy. Alchemists aspired to achieve extraordinary and, by our standards, *impossible* transformations: the transmutation of base metals into gold, the prolongation of human life, and the attainment of celestial perfection even within the flawed elemental world. They justified these ends with reference to a variety of ideas and explanations, ranging from analogies with the natural world to comprehensive physical theories. Since a variety of explanations and approaches were available, they sometimes had to choose between them. Even though it is not obvious that we can discuss alchemical ideas within the same framework as that used for modern scientific theories, the fact that alchemical explanations were not given arbitrarily nevertheless raises some interesting questions about what "theory choice" involves.

I shall begin with two caveats: one to do with medieval science, and one to do with theories.

Edward Grant has described medieval science as "empiricism without observation" [1, 2]. Its principles were discussed and refined without necessarily being subjected to empirical testing. This is not to say that medieval people had no science, nor that they were incapable of amending explanatory frameworks in order to accommodate observations. However, it does mean that received wisdom from authoritative sources generally carried far more weight than it does today, even when flawed. Rather than speaking of "science," I shall therefore use the contemporary scholastic category of "natural philosophy" to describe the medieval pursuit of natural knowledge.

Natural philosophy was concerned with events in the terrestrial world: a sphere composed of the four Aristotelian elements of earth, air, fire and water. In that

Earlier versions of this paper were delivered at the Summer Symposium of the International Society for the Philosophy of Chemistry (Leuven, September 2012) and at the 7th Integrated HPS Workshop (UCL, June 2012). I am grateful to my interlocutors at these events, and to members of the AD HOC Reading Group, London, for their helpful comments.

J.M. Rampling (✉)
Department of History, Princeton University, Princeton, NJ 08544-1017, USA
e-mail: rampling@princeton.edu

© Springer International Publishing Switzerland 2016
E. Tobin and C. Ambrosio (eds.), *Theory Choice in the History
of Chemical Practices*, SpringerBriefs in History of Chemistry,
DOI 10.1007/978-3-319-29893-1_2

world, change was explained in terms of the imposition of new specific forms onto a basic material substrate—for instance, when good wine became corrupt or sour, the form of the wine was replaced by that of vinegar. Ideas about alchemical transmutation were also expressed in terms of matter and form, even if these explanations were not set out with the level of detail and rigour that modern scientists demand of their theories.

The second caveat concerns my use of "theory." Alchemical texts are often (although not always) divided into two parts: *theorica* and *practica*. The practica usually provides an ensemble of recipes for a range of alchemical products, including transmutational elixirs, medicines, blanchers, pigments, and artificial gems. The theorica generally comes first, providing an introduction to the operative section of the treatise, and outlining some fundamental systems and explanations. The theorica might, for instance, discuss the system of Aristotelian elements and the relationship between matter and form; the manner by which metals are generated in the earth; the composition of the various metals and their characteristic properties; and so on. This introduction, often influenced by Aristotelian natural philosophy, provides the basic understanding of metallic substances necessary to support the next level of explanation: namely how one kind of metal may be transformed into another.[1]

A present day scientist might object that such understanding would be better described as a *world view*, or a *cosmology*, or a *framework*, rather than a theory or hypothesis in the modern sense. For instance, while medieval authors often suggest explanations for alchemical transmutation, the "truth" of such explanations may seem to be assumed, rather than supported by clear, testable, replicable hypotheses. Nevertheless, these explanations do reveal a real engagement with contemporary natural philosophy and with empirical results. To avoid ambiguity, I shall adopt an archaic English term, "theorick," when referring to the explanations of alchemical change proposed by medieval alchemists.

Which brings us to alchemy itself. Clearly, in the context of theory choice, alchemy poses particular problems. How can we discuss the falsification of an approach that we already *know*, from our privileged historical vantage point, could not have worked? In the case of alchemical transmutation, the very phenomenon that each theorick purports to explain is known to be impossible—or at least, impossible using the techniques available to medieval practitioners. Is it even meaningful to assess alchemical theories using the standards suggested by Kuhn [3] —of accuracy, consistency, broad scope, simplicity, and fruitfulness?

Alchemists were, in fact, deeply concerned with the rationality of their pursuit. Although widely practised, alchemy was never formally part of the university curriculum, so its practitioners sought alternative ways of legitimising it as a subject worthy of serious intellectual consideration [4]. Historians of science have therefore found it fruitful to concentrate on demarcation problems: asking how theories of transmutation arose in response to critique from medieval sceptics. These scholars

---

[1]For a more detailed introduction to alchemical ideas, see Lawrence M. Principe, *The Secrets of Alchemy* (Chicago: University of Chicago Press, 2012).

have sought to show that alchemical ideas were rational according to the standards of medieval natural philosophy. Indeed, William Newman has gone so far as to argue that alchemy was actually *more* empirically grounded, and hence capable of offering better explanations of observable phenomena, than standard Aristotelian physics [5]:

> The alchemists of the High Middle Ages established an experimentally based corpuscular theory that would develop over the course of several centuries and eventually supply important components to the mechanical philosophy of the Scientific Revolution.

Yet this "corpuscular theory" is only one (albeit an influential one) among several kinds of explanation encountered in medieval alchemy, and practising alchemists did not always agree. This raises a new question: why did alchemy's supporters opt for one explanation rather than another? Were they intrested in explanatory power, empirical evidence, or consistency with other natural philosophical doctrines (including the works of past adepts)? Rather than asking how alchemists attempted to justify their art to sceptics, I shall investigate alchemists' engagement with *one another*, by considering some basic disagreements between different views of alchemical transmutation.

2.1 The Nature and Genesis of Metals: Competing Views

I shall start by outlining three different views on the nature and generation of metals, which all rely on the same basic idea—usually referred to as "Sulphur-Mercury theory."

Medieval views on the structure of metals were influenced by a tradition originating in Aristotle's *Meteorology*, and subsequently developed by medieval Islamic authors, including the semi-legendary alchemist Jābir ibn Hayyān, and the Persian polymath Ibn-Sina, or Avicenna.[2] According to this view, metals were formed within the earth by the commixtion of two vapours, or exhalations. One, a moist, smoky vapour, was described as "Mercury" —not elemental quicksilver, but a principle of moistness and fluidity. The second principle, a dry, earthy exhalation, was called "Sulphur." The composition and properties of different metals were explained in terms of their respective proportions of Sulphur and Mercury. Thus quicksilver is runny because it consists almost entirely of the Mercury principle. However, it does not wet the hands, since its small component of Sulphur imparts dryness to its surface. Conversely, iron's high melting point suggests a high Sulphur content.

The proportion and purity of the two principles also determine the quality of the resulting metal. The Mercury and Sulphur in lead are corrupt, imparting a dark colour. Only gold has the optimum proportion of clean Sulphur and Mercury. As evidence for its perfection, gold is able to retain its form: it does not tarnish, and is

---

[2]The theory is fully discussed by Norris [6]. On some aspects of its medieval reception, see Newman [5, 7].

only with difficulty persuaded to join with other substances (an important exception being quicksilver). The theory also explains why metals are found in different states of purity, depending on the relative proportion of the principles.

This idea has some empirical support. For instance, calxes are made when a "body" (or metal) loses some of its "humidity": the moist quality which makes metals fusible. Calxes are generally not fusible: they have a dry and sometimes brittle appearance, like stone, crystal, or powder. The Sulphur-Mercury theory also helps explain the apparent retrieval of chemical substances following their dissolution in corrosives or compounding with other substances—a reconstitution that made no sense in terms of conventional Aristotelian physics, which stresses that a substance, once it has lost its original form (for instance through dissolution in an acid) cannot readily regain it. As Newman has shown, this explanation allows for an intermediate state between the metals and their most basic building blocks, the four Aristotelian elements (earth, air, fire and water). Rather than being reduced to these elements and losing its form, a metal need only be reduced to its constituent Sulphur and Mercury, principles that may then be recombined to produce the "lost" metal, in a process later described as "reduction to the pristine state" [5].

However, this explanation raised fresh problems. For instance, does each kind of metal have its own specific form, or should differences between metals be viewed merely as superficial variations, or "accidents," of a single, basic, metallic species? Unfortunately, Aristotle had neglected to provide detailed instruction on this matter. The great thirteenth-century Dominican thinker and Aristotelian commentator, Albertus Magnus (1193–1280), therefore undertook to produce his own book on minerals, the *Liber Mineralium*.[3] In it, he argued that each metal had its own specific form; thus, lead was substantially different to gold. On these grounds, Albertus stated his view that alchemical transmutation, although possible, must be very difficult to attain, since it entailed the imposition of a new form—that of gold—upon a different substance [12]:

> And alchemy also proceeds in this way, that is, destroying one substance by removing its specific form, and with the help of what is in the material producing the specific form of another [substance]. And this is because, of all the operations of alchemy, the best is that which begins in the same way as nature, for instance with the cleansing of sulphur by boiling and sublimation, and the cleansing of quicksilver, and the thorough mixing of these with the material of metal; for in these, by their powers, the specific form of every metal is induced.

In setting out this view, Albertus explicitly distanced himself from another position which, he said, was common among alchemists, particularly one "Callisthenes".[4] The alchemists' position, as Albertus characterises it, supposes that one metal can transform into another one through a natural process of digestion and maturation beneath the earth. The least perfect metals gradually ripen into the more

---

[3]Published in English translation as Albertus Magnus, *Book of Minerals*, translated by Wyckoff [8]. On Albertus' alchemy, see also Partington [9], Kibre [10], and Halleux [11]. A large number of alchemical tracts were later pseudonymously attributed to Albertus.

[4]As Wyckoff notes, this name is a mistake for Khalid ibn Yazid, one of the protagonists of the early *Liber de compositione alchimiae*, and supposed author of the *Liber trium verborum* [13].

perfect ones, even lead eventually becoming gold. Once gold is attained, the process stops, nature having achieved her ends.

This process causes some difficulties for the devout Aristotelian, by implying that the specific form of metal is gold, and that the lesser metals are merely faulty versions of this perfect substance. It also enabled alchemists to argue that, using artificial processes, they could therefore help nature along, but in shorter time [13]:

> For they seem to say that the specific form of gold is the sole form of metals, and that every other metal is incomplete—that is, it is on the way towards the specific form of gold, just as anything incomplete is on the way towards perfection. And for this reason metals which in their material have not the form of gold must be 'diseased'; and [the alchemists] try to find a medicine which they call elixir, by means of which they may remove the diseases of metals... and thus they speak of 'bringing out' the specific form of gold.

Within the overarching Sulphur-Mercury framework, we can therefore identify at least three distinct explanations, or "theoricks," for alchemical transmutation.

First, in his *Book of the Remedy* (sometimes considered to be part of Aristotle's *Meteorology*), the Persian polymath Avicenna argued that each metal belongs to a distinct species, and it is not possible to transmute one species into another. The alchemist would first have to break matter down into its constituent elements, and then reconstruct the new species from scratch—which cannot be done through art.

Second, we have Albertus' position in the *Liber Mineralium*. Albertus interprets Avicenna's "species" differently (and probably not quite accurately), as *specific form* [7]. Each metal has its own specific form, which must be destroyed in order for nature to replace it with a new and better one. If alchemists can find a way of stripping a metal of its original specific form, it may be possible to transmute one metal into another, albeit with great difficulty.

Third is the "maturation" approach that Albertus attributes to Callisthenes. This views all the metals as part of a continuum, within which the less perfect metals are gradually "perfected" until they become gold. By inference, gold is the specific form of all metals. Not only is transmutation possible, but it actually forms part of the natural evolution of metals. Albertus criticises this approach, since the notion of gold as the specific form of metals seems to trouble him.[5]

While each of these theoricks is underdetermined by evidence, they are not without empirical support. For instance, Albertus' theorick of "specific form" can explain why metals are only discovered in discrete species, rather than in a halfway state between metals. It is also broadly consistent with the kind of natural philosophy being discussed in the medieval universities, and hence could be seen as compatible with the "normal science" of thirteenth-century Europe. On the other hand, the "maturation" theorick explains why different ores are often found within

---

[5]"For there is no reason why the material in any natural thing should be stable in nature, if it were not perfected by a substantial form. But we see that silver is stable, and tin, and likewise other metals; and therefore they seem to be perfected by substantial forms... And as to the experiments which [the alchemists] bring forward, not enough proof is offered." [14].

the same mines—the metals co-exist because they are continually transforming into one another. Interestingly, Albertus himself reports "making long journeys to mining districts, so that [he] could learn by observation the nature of metals" [15].

2.2 Vegetable or Mineral?

As a theory of metallogenesis, Sulphur-Mercury meets some of Kuhn's criteria in terms of accuracy, broad scope, and so forth. As explanations of transmutation, however, the various theoricks derived from it leave a lot to be desired. Each uses ideas about the formation of metals as the basis for arguing whether or not transmutation is possible. They say nothing about how transmutation is to be achieved, except in the vaguest terms—the implication being that the alchemist must somehow replicate and abbreviate a series of processes that would normally occur naturally over thousands of years in the bowels of the earth.

Each explanation also applies only to metallic bodies. Yet to do really interesting chemistry, metals are not enough. For instance, to dissolve metals and their calxes requires solvents made from non-metallic ingredients. *Aqua fortis*, *aqua regia*, and other mineral acids were made using vitriol (commonly regarded as a spirit rather than a metallic body) and salts. Distilled vinegar was of course derived from wine, a "vegetable" product. Other salts were obtained from "animal" substances like hair, urine or eggshells.

Yet, from the perspective of medieval natural philosophy, transmutation is much easier to explain if discussion is confined to the metals and their constituent principles. Alchemical authors are often very critical of reliance on other minerals, including salts, vitriols, and alums, yet are nevertheless forced to recognise their importance. In the words of ps.-Geber [16]:

> Because we see adherence to the bodies accompanied by alteration to occur in no other material but the spirits, we cannot therefore be freed from their use, nor may we escape their preparation by cleaning, which is accomplished by sublimation.[6]

The real ire of alchemists like ps.-Geber, however, was reserved for those who sought to use animal or vegetable ingredients in their work. To scholastic authors it was obvious that, in order to create gold, one must start with some kind of metal. Paradoxically, this argument was supported using an analogy with animal generation: just as man begets man and beast begets beast, so metals may only be generated from metals.

In fact the debate over organic products was not new in the thirteenth-century West. A tenth-century Arabic treatise, the *Mā 'al-waraqī* of Muhammed Ibn Umayl, attacked alchemists who, like the polymath Abu Bakr Muhammad ibn Zakarīyā al-Rāzī ("Rhazes" to the Latins), recommended such animal ingredients as

---

[6]Note that "bodies" here denote metals; while "spirits" include alums and salts which do not remain fixed in fire.

hair and eggs [17, 18]. Yet these ingredients often produce chemically interesting results. The problem came to a head in the mid-fourteenth century, with the development of an entirely new approach to alchemical transmutation, relying on alcohol as its primary ingredient.

2.3 Inconsistent Results

From the end of the thirteenth century, alchemists embraced an exciting new technology: distillation. Repeated distillation of spirit of wine was found to yield a liquid with peculiar properties: clear as water, yet highly flammable. A drop would burn the tongue. A piece of meat or vegetable matter placed in it would not decay. It could be used as a solvent for oils and other substances that did not dissolve in water.

These qualities led alchemists to infer that the "quintessence of wine" should be able to preserve living human bodies as well as it preserved meat. Unlike mineral acids, it was also a safe solvent for human ingestion. From this arose the notion that quintessence could be used both as an alchemical medicine for internal use and as an ingredient in the transmutation of metals. This theory was set out in great detail in one of the most influential alchemical treatises of the fourteenth century, the *Liber de secretis naturae, seu de quinta essentia* (The Book of the Secrets of Nature, or, concerning the Quintessence), attributed to the Majorcan philosopher Raymond Lull (ca. 1232-ca. 1316).[7] Pseudo-Raymond described the quintessence as a "vegetable mercury," or "resolutive menstruum" [23].

However, although absolute alcohol is indeed effective for extracting plant essences, its effectiveness is limited. In the late fifteenth century, the English alchemist George Ripley rejected quintessence of wine of the kind described by Raymond in the *Liber de secretis*. Apparently on the basis of his own experience, he observed that even multiple distillations fail to produce a quintessence sharp enough to dissolve metallic calxes:

> Some assert that this fire is a water drawn from wine, according to the common way, and should be rectified, being distilled as many times as possible... yet, when water of this kind (which fools call the pure spirit), even if rectified a hundred times, is put upon the calx of whatever body, however well prepared, nevertheless we see it will be found weak and entirely insufficient for the act of dissolving our body with conservation of its form and species. Wherefore it seems there is an error in the choice of this principle, which is called the resolutive menstruum.[8]

---

[7]"Raymond" had in fact borrowed the concept, and most of the text, from John of Rupescissa's *Liber de consideratione quintae essentiae* of 1351-52. On John of Rupescissa, see Taylor [19], Multhauf [20], Halleux [21], and DeVun [22].

[8]"Quidam autumant ignem istum aquam esse | a vino tractam vulgari modo rectificarique debere eam multotiens distillando vt possit | ab ea eius aquosum flegma vires et potentias sue igneitatis impediens, penitus | extirpari. Sed cum talis aqua centies rectificata quam dicunt fatui spiritum esse | purum mittitur super calcem corporis optime preparatam: videmus quod ad actum dissoluendi | corpus cum conseruacione sue forme et speciei impotens ac omnino insufficiens reperitur | Quare videtur quod in electione huius principij quod menstruum resolutiuum dicitur | error sit" [24]. My

Ripley also knew that Raymond's advice here conflicted with his instructions in another text, in which he recommended the use of mineral rather than vegetable substances: "If, as Raymond says, the resolutive menstruum springs from wine or the tartar thereof, how is what the same philosopher says to be understood: 'Our water is a metalline water, because it is produced from a metalline kind'"?[9]

Ripley would not have phrased the problem in quite these terms, but his basic concern is with the accuracy and consistency of his source. He faced an additional difficulty, namely the fact that an important authority seemed to be contradicting himself. Unknown to Ripley, there was a good reason for this: the many works attributed to Lull were all pseudepigraphic, and written by various authors, presumably engaged in different (although related) types of practice. Fortunately for Ripley, one of these seemed to offer a compromise: it described a solvent made using distilled vinegar, which was more penetrating than the spirit of wine. By dissolving lead salts in the vinegar and then distilling the resulting product, Ripley obtained a solvent that, he believed, was both metalline *and* vegetable in its nature. This substance fulfilled the requirements of all Ripley's sources, preserving Raymond's authority—and, crucially, justifying Ripley's faith in his own observation.

2.4 Conclusion

Medieval alchemists had a difficult task before them, in more ways than one. They had to adjudicate between earlier theories and explanations, because their choice of explanation would often determine the starting matter, processes, and ends of their practice. Since no one had ever witnessed a successful transmutation, they also had to be guided by textual authority to a high degree: resulting in the kind of exegetical shenanigans that have just been seen in the case of George Ripley and the contradictory contents of the pseudo-Lullian corpus.

This still leaves the question of the extent to which practice was genuinely guided by theory. The textual evidence suggests that theories were often amended in light of empirical observations, albeit not in ways that we are used to in the modern world. Rather than replacing the earlier theory, alchemical practitioners had a vested interest in preserving their predecessors' authority. New or revised texts seldom contradicted the revered authority of a past adept. Instead, alchemists preserved the authority of past sources (and demonstrated their own authority in the process), by reinterpreting earlier works in such a way that they gave support to the

---

(Footnote 8 continued)

transcription and translation. Italics here denote the expansion of abbreviated text. For more detailed discussion of this passage, see [25].

[9]"Sed si a vino oritur menstruu*m* resolutiuu*m* vt vult Ray*mundu*s vel a | tartaro eius: quomo*d*o intelligitur q*uod* ide*m* p*hilosoph*us dicit. Aqua n*ost*ra est aqua | metallina, quia ex solo genere metallico generatur" [24].

desired findings or approach. Straw men could then be sacrificed in their place: those unsuccessful alchemists, or "fools," who had misunderstood the true meaning of the philosophers.

For alchemists, choice between "theoricks" seems to have been based in part on the compatibility between authority and practice. Authority, however, could still be interpreted in light of practice, while practical programmes might in turn be shaped by the instructions and expectations of authorities. Alchemists did not choose ingredients and processes at random, but allowed their practice to be shaped by the records of their apparently successful predecessors. Yet experience also forced changes, including the adoption of organic substances like vinegar, even when such modifications seemed to conflict with established views.

I have presented this medieval tussle as a case study for an integrated approach to history and philosophy of science. Other than the corpuscular approach championed by William Newman, specific alchemical "theoricks" have received little attention from either philosophers or historians of science. The reasons given by alchemists for selecting one or other approach remain understudied. Yet their complexities and difficulties provide intriguing counter examples to more familiar examples taken from medieval mathematics, astronomy, or optics. Studying alchemical texts will not teach us how to generate gold, except in the most metaphorical sense. It may, however, offer insight into the processes by which pre-modern practitioners devised and used experiments to extend their knowledge of natural processes—a goal that is as much the historians' as the philosophers' stone.

References

1. Grant E (2002) Medieval natural philosophy: empiricism without observation. In: Leiijenhorst C, Lüthy C, Thijssen JMMH (eds) The dynamics of Aristotelian natural philosophy from antiquity to the seventeenth century. Brill, Leiden, pp 141–168
2. Grant E (2010) The nature of natural philosophy in the Middle Ages. Catholic University of America Press, Washington, DC, Chapter 8
3. Kuhn T (1972) Objectivity, value judgment, and theory choice. In: The essential tension. University of Chicago Press, Chicago, pp 320–339
4. Obrist B (ed and trans) (1990) Constantine of Pisa. The book of the secrets of alchemy: introduction, critical edition, translation and commentary. Brill, Leiden
5. Newman WR (2006) Atoms and alchemy: chymistry and the experimental origins of the scientific revolution. University of Chicago Press, Chicago, p 26
6. Norris JA (2006) The mineral exhalation theory of metallogenesis in pre-modern mineral science. Ambix 53:43–65
7. Newman WR (1989) Technology and alchemical debate in the late Middle Ages. Isis 80:423–445
8. Albertus Magnus (1967) Book of minerals. Wyckoff D (trans), Clarendon Press, London
9. Partington JR (1937) Albertus Magnus on alchemy. Ambix 1:3–20
10. Kibre P (1980) Albertus Magnus on alchemy. In: Weisheipl JA (ed) Albertus Magnus and the sciences: commemorative essays 1980. Pontifical Institute of Mediaeval Studies, Toronto, pp 187–202

11. Halleux R (1982) Albert le grand et l'alchimie. Revue des Sciences Philosophiques et Theologiques. 66:57–80
12. Albertus Magnus (1967) Book of minerals. Wyckoff D (trans), Clarendon Press, London, Book 3, p 179
13. Albertus Magnus (1967) Book of minerals. Wyckoff D (trans), Clarendon Press, London, Book 3, p 171
14. Albertus Magnus (1967) Book of minerals. Wyckoff D (trans), Clarendon Press, London, Book 3, p 173
15. Albertus Magnus (1967) Book of minerals. Wyckoff D (trans), Clarendon Press, London, Book 3, p 153
16. Newman WR (ed and trans) (1991) The Summa Perfectionis Magisterii of Pseudo-Geber: a critical edition, Translation and Study. Brill, Leiden, p 682
17. Stapleton HE (ed) (1933) Three Arabic Treatises on Alchemy by Muhammad bin Umail (10th Century A.D.). Memoirs of the Asiatic Society of Bengal. vol. XII, Calcutta, p. 142
18. Stapleton HE, Lewis GL, and Taylor FS (1949) The Sayings of Hermes Quoted in the Mā' al-Waraqī of Ibn Umail. Ambix 3:69–90
19. Taylor FS (1953) The idea of the quintessence. In: Underwood EA (ed) Science, medicine and history, 2 volumes. Vol 1, Clarendon Press, Oxford, pp 247–265
20. Multhauf RP (1954) John of Rupescissa and the origin of medical chemistry. Isis 45:359–367
21. Halleux R (1981) Les ouvrages alchimiques de Jean de Rupescissa. Histoire littéraire de la France 41:241–277
22. DeVun L (2009) Prophecy, alchemy, and the end of time: John of Rupescissa in medieval Europe. Columbia University Press, New York
23. Pereira M (2002) Vegetare seu transmutare. The vegetable soul and pseudo-Lullian alchemy. In: Domínguez RF, Villalba-Varneda P, Walter P (eds) Arbor scientiae: der Baum des Wissens von Ramon Lull. Akten des Internationalen Kongresses aus Anlaβ des 40-Jährigen Jubiläums des Raimundus-Lullus-Instituts der Universität Freiburg. 29. September—2. Oktober 1996. Brepols, Turnhout
24. Ripley G Medulla alchimiae. In: Cambridge, Trinity College Library MS R.14.58, Part 3, folio 5r
25. Rampling JM (2014) Transmuting sericon: alchemy as practical exegesis in early modern England, Osiris 29:19–34

Chapter 3
Theory Choice in the Seventeenth Century: Robert Boyle Against the Paracelsian *Tria Prima*

Georgiana D. Hedesan

Robert Boyle's famous *Sceptical Chymist* (1661) is a dialogue on matter theory, between a Peripatetic Aristotelian (Themistius), a Chymist (Philoponus) and a Sceptic (Carneades), and moderated by a supposedly impartial individual (Eleutherius). At first glance, the book seems to offer an ideal case in which to study theory choice. Boyle introduces at least three types of competing theories of matter: the four-element theory of Aristotle, the three-principle theory of Paracelsus and the atomistic theory. However, the dialogue quickly degenerates as the Aristotelian and the Chymist strangely disappear from the picture and the Sceptic and his moderator remain to talk amongst themselves [1]. Thus, what had promised to be an unbiased debate of the virtues of the three theories turns into a discussion on one system only, the Paracelsian theory of the 'three principles' of Sulphur, Mercury and Salt, also referred to as the *tria prima*. The promise of theory choice between three competing systems hence thwarted, a more fruitful avenue of research is to analyse the book from the point of view of a mitigated type of theory choice, that between acceptance of the *tria prima* theory or its rejection and exploration of other theories. The questions connected to it are the following: (1) is Boyle offering a true choice to the reader amongst these two options? (2) If not, does Boyle offer sufficient proof to sway an impartial reader to his choice? (3) Finally, what does Boyle's argument tell about his own choice of theory?

In fact, question (1) can be quickly answered. The dialogue is dominated by one viewpoint only, that of the Sceptic Carneades (a Boyle alter-ego), who begins the dialogue already convinced of the inferior status of the Paracelsian *tria prima*. His interlocutor, Eleutherius, does not really challenge his views. Hence the book should be seen as an attempt to influence the reader toward rejecting the *tria prima*. This is evident in the fact that the Paracelsian Chymist invited to the dialogue

G.D. Hedesan (✉)
History Faculty, University of Oxford,
George Street, Oxford OX1 2RL, UK
e-mail: georgiana.hedesan@history.ox.ac.uk

© Springer International Publishing Switzerland 2016
E. Tobin and C. Ambrosio (eds.), *Theory Choice in the History
of Chemical Practices*, SpringerBriefs in History of Chemistry,
DOI 10.1007/978-3-319-29893-1_3

hardly speaks at all throughout the book, while Carneades takes an explicit anti-Paracelsian stance. We should not expect to find a fair hearing of the *tria prima* in this book. Instead, we should read it as a justification for Boyle's own choice to reject the Paracelsian theory.

Given this immediate conclusion, we are automatically referred to question no. (2): does Boyle offer sufficient proof to sway an impartial reader to reject the *tria prima*? To answer the question, we shall first review the theory of the *tria prima* and its status in Boyle's time.

3.1 The *Tria Prima* and Its Supporters

The *tria prima* was the brainchild of the Swiss physician, philosopher and alchemist Theophrastus von Hohenheim, best known as Paracelsus (1493–1541). According to Paracelsus's mature theory, all matter was comprised of Salt, Sulphur and Mercury [2]. His idea was clearly an extension and augmentation of the medieval alchemical theory of Sulphur and Mercury, which has been referred to in the previous chapter.[1] Paracelsus modified the Sulphur-Mercury theory in two major ways: by adding a third principle, Salt, and by expanding it beyond metals. The three principles came to constitute the building blocks of all beings in the universe.

Paracelsus was not always clear of what he meant by the three principles. However, his view of them possessed several traits that later followers agreed with. One was that the *tria prima* was constitutive of all bodies, albeit invisibly. They were active in creating the body, and remained active even when isolated from a corpse. Hence they played a very important role in a medical context: the skilful alchemist could extract these active principles from the body and 'inject' them into powerful medicines.

While Paracelsus is keen to emphasise that the *tria prima* operate within the living body as occult forces [3], he believed that there was a way to make the *tria prima* visible, hence to demonstrate their existence. The method he proposed was that of the fire [4]: 'the fire proves the three substances and presents them pristine and clear, pure and clean.'

Paracelsus had a peculiar view of fire which determined his belief in its power of rendering the invisible manifest. Fire, he says in his mature works, is heaven (*coelum*), and not an element [5]. He viewed fire as being something more spiritual and active than the sublunary elements, and its primary role was that of separating bodies. In Paracelsian philosophy, separation (*Scheidung*) was a fundamental act whereby the invisible becomes visible, and the spiritual material [6]. Yet this act also corresponded to destruction, because the separation of a body by fire meant its death. By applying fire to a body, one could extract its fleeting spirits and incorporate them in medicine.

---

[1]See previous chapter by Dr. Jennifer Rampling.

As 'proof' of the *tria prima*, Paracelsus used the example of green wood, which when burnt gives out flame (the 'Sulphur'), smoke ('Mercury') and ash ('Salt'). This example was used by other Paracelsian followers as well [7]. Although called an 'experiment' by Boyle, it was a very simplistic and not especially alchemical example; rather, it was more of a rhetorical device to clarify the theory and capture people's imagination [8]. Moreover, it was not a very solid or original argument against the Aristotelian philosophy, which used the same experiment to 'prove' the four elements: earth (ash), water (sap), air (smoke) and fire [7]. Indeed, a supporter of alchemy like Daniel Sennert found that the wood example was neither relevant nor chemically correct [9].

Still, the 'burning wood' example could inspire much more sophisticated experiments. A much more alchemical and potent method of revealing the *tria prima* was through distillation. Thus, the organic distillation of oak chips yielded five fractions, out of which the active ones were deemed to be Mercury, Sulphur and Salt [10].

The *tria prima* provided a valuable theoretical framework for alchemists, and satisfied their need to understand the diverse, and often strange phenomena they experienced in the laboratory. It was also a theory that appealed to the senses: the *tria prima* could be experienced directly, without mitigation, a fact that alchemists often boasted about [11].

We can further understand the appeal of the theory of the *tria prima* by using Thomas Kuhn's arguments for rational theory choice as expressed in 'Objectivity, Value Judgment and Theory Choice' [12]. Thus, we can admit that, in the historical context, the *tria prima* was roughly accurate (in agreement with 'existing experiments and observations'), broad in scope (applied to all bodies in the universe), simple (in the sense used by Kuhn of ordering confusing phenomena) and generally fruitful (it certainly led to more, rather than less, chemical inquiry and was strongly inclined to the value of useful knowledge).

However, there were downsides to this theory. One problem can be related to Kuhn's criterion two, consistency: the three-principle system was a 'local' theory, popular and useful only in medical and alchemical practice, and fundamentally different from the larger framework of natural philosophy. Even alchemists themselves often tended to look at the *tria prima* pragmatically, rather than philosophically. Another problem was related to Kuhn's criterion one, accuracy, and referred to the degree of experimental precision: alchemists were not able to obtain the 'pure' principles in the laboratory, as there were always traces of the other principles in them [13]. Thirdly, the *tria prima* was not a standardised account. Variations on the subject seemed endless [14].

Generally, one can affirm that it was an influential theory in its field, and it was hence unsurprising that it won many alchemical practitioners by the beginning of the seventeenth century. This was, in fact, the 'golden age' of the *tria prima*, when numerous alchemical writers presented the theory as unquestionable fact [15–17]. Arguably the most influential supporter of the *tria prima* of this period was Joseph Du Chesne, or Quercetanus (1544–1609), whom Boyle recognised as 'the grand stickler for the *Tria Prima*' [18]. Du Chesne proposed further evidence for the theory, emphasising that the *tria prima* can be extracted from marine salt, vitriol, common sulphur and others. He suggested the theory that the quantity of the principles in each metallic body is different, hence copper has a large quantity of Sulphur, less so of vitriolic salt, and least of all Mercury [19]. In line with such argument, Du Chesne argued that gold has an equal part of all the three principles, a matter which makes it the most noble of metals, and also the hardest to analyse in its components [19].

Du Chesne's insistence on the *tria prima* had a major influence on the development and popularity of this doctrine in alchemical circles and beyond. Yet, paradoxically, Du Chesne was also the originator of a competing theory. In an alchemical poem called *The Great Mirror of the World*, he affirmed that, as far as the distillation process was concerned, the distinction medical alchemists made between the five fractions was meaningless [10]. By the 1620s, a new generation used this opinion to advance a new theory of the 'five elements'. Étienne de Clave and Antoine Villon strongly believed that the division of the fractions into active and passive had to be abandoned [10]. The five-element theory weakened the theory of *tria prima* since de Clave and others advanced their teaching by attacking the 'obsolete' three principles of the Paracelsians. Their argument was clearly mirrored in Boyle's attack of the *tria prima*, where he rejects the Paracelsian division of active principles from passive elements [20].

3.2 Rising Dissent: The Role of Van Helmont

As the case of the five-element theory suggests, the threat to the theory of the *tria prima* did not come from the outside, but from inside alchemical practice. By comparison, the Scholastic resistance to the *tria prima* did not seriously affect it. Perhaps typically of chemists, Boyle gives to Peripatetic arguments short thrift; he was not prepared to accept a theory that was based on rhetoric rather than experimentation.

Clearly, the real blow to the *tria prima* could only have come from the alchemical community itself. In the 1640s, the strongest challenge to the three principle theory came from the Flemish physician and alchemist Jan Baptist Van Helmont (1579–1644).

Van Helmont's rejection of the theory stemmed from two main sources. One was his radical view of matter as being a composite of water impregnated by invisible seeds. The idea that all matter was essentially 'water' (or a watery substance) was a rather popular alchemical view in the period and was often linked to interpretations of the Genesis [21, 22]. In Van Helmont's system, the *tria prima* could not be the fundamental components of bodies, since they too could be decomposed into water by such universal solvents as the mysterious Alkahest.

The other source of dissent had to do with Van Helmont's non-Paracelsian view of the role of fire in chemical work. As I already suggested above, Paracelsus viewed fire as the ideal instrument to analyse bodies. However, anti-Paracelsian thinkers like Thomas Erastus, Jean Riolan and Marin Mersenne disagreed, arguing that fire did not separate, but rather compounded other bodies [9, 23]. This view implied that the *tria prima* were not 'first principles' of matter, but simply by-products of the action of fire.

It is not clear how Van Helmont came by this non-Paracelsian view, yet it resonated well with his belief that dissolution of bodies can best be achieved by means of solvents rather than fire. Van Helmont argued that fire did not analyse bodies, but was able to produce new ones [24].

Such a view of fire left no room for the *tria prima* as constituent principles of bodies. If fire compounded, rather than separated bodies, the phenomena associated with the *tria prima* had no meaning in terms of understanding the structure of matter. Van Helmont did not deny the importance of the three principles in medicine, but he certainly negated the *tria prima*'s role in matter theory.

To buttress his arguments against the *tria prima*, Van Helmont used experimental evidence. He argued that there were several bodies out of which the fire could only produce one or two of the *tria prima* or none at all [24]. Such bodies, he pointed out, were primarily gold and mercury, but also sand, flint and stones that do not contain lime [25]. He also maintained that water, one of his two primordial elements, could not be further reduced into the Paracelsian *tria prima* [26]. He also criticised the Paracelsian assumption that the salt obtained from urine was one of the *tria prima*, when in fact it was only salt water that had not been truly separated into its components [27].

Although such 'experimental' criticism is important, it cannot obscure the fact that for Van Helmont it was his views of the nature of compound bodies, and of the role of fire, that determined his views of the *tria prima*. In his case, theory determined practice, rather than vice versa. Van Helmont never hid the fact that he considered that knowledge came from above rather than below. Experiment played an important epistemological role in Van Helmont, albeit not a determinant one in the way we might expect from modern scientists.

3.3 Boyle's Helmontian Rejection of the *Tria Prima*

It is no secret that Boyle, at least in his youth, was a fervent Helmontian. Newman and Principe have emphasised Boyle's indebtedness in chemical practice to the strong Helmontian supporter, George Starkey [28]. Whether together with Starkey or on his own, Boyle is known to have prepared a number of Helmontian remedies [29]. His early writings, including *The Usefulnesse of Experimental Philosophy* and the 'Reflexions on the experiments vulgarly alledged to evince the 4 peripatetique elements, or ye 3 chymicall principles of mixt bodies' (both written prior to 1660), evince strong Helmontian influences [30]. This early interest and practice could make the reader at least slightly suspicious as to his objectivity toward the Paracelsian *tria prima*.

In the *Sceptical Chymist*, Boyle did not try very hard to hide his preference for Van Helmont's work over that of Scholastics and Paracelsians. His praise of Van Helmont rather contrasts with the 'sceptical' attitude of Carneades, revealing his fundamental bias for Helmontian thought, while rejecting the title of 'Helmontian'.[2] Indeed, Carneades held mostly positive views of the Flemish alchemist and even used Helmontian terms in his speech [33].

Indeed, the work shows that Boyle was enthralled with many Helmontian experiments, particularly that of the willow tree, which was supposed to show that all vegetables turn into water. His alter-ego 'Carneades' carried out similar experiments that confirmed Van Helmont's observations. He also paid close attention to Van Helmont's account of a wondrous solvent, the Alkahest, which was supposed to dissolve all bodies into water. Although Carneades admitted that he had not made the Alkahest, his interest in the subject suggests that he was inclined to believe in it [34]. As pointed out above, the acceptance of the Helmontian view that there is a universal solvent that could transform everything into elementary water implied an automatic rejection of the *tria prima*.

A barely veiled adherence to Van Helmont's doctrine is also evident in Boyle's repetition of the Flemish physician's views of the role of fire.[3] Fire, Carneades notes, 'produces Concretes of a new indeed, but yet of a compound Nature' [36]. At least in some cases, fire does not cause separation, but a union that cannot be broken up by fire itself [37]. Carneades also affirms that other instruments, such as solvents, have an ability to separate substances, hence agreeing with Van Helmont's preference for such means of analysing bodies [38].

This adherence to Helmontianism does not mean that Boyle was blindly reproducing Van Helmont's assertions without trying any of them out. In several cases it is clear, or at least likely that he repeated or even improved upon the experiments advanced by Van Helmont, usually with very similar results. Yet we must also understand that his work was carried out within the Helmontian framework. The choice of experiments (such as the reduction of vegetables into water) was strongly dictated by this background.

One may legitimately ask why, in this case, Carneades (and implicitly Boyle) rejected the label of 'Helmontian'. This was probably so because the 'Helmontians' did not always have a positive connotation in polite English circles: as Clericuzio [30] and Debus [39] have pointed out, Boyle distanced himself from such vociferous Helmontians as John Webster, who rejected the traditional university curricula. Moreover, Boyle was also keen to affirm his own authority.

Nevertheless, Carneades' divergence from Van Helmont was not very substantial. He largely agreed with Van Helmont's principle that all bodies sprang out of water, although he did not think water was the ultimate element [40]. He also

---

[2]See, for instance [31]. For his protest at being called a Helmontian, see [32]. This defence suggests that at least some persons saw him as such.

[3]Boyle acknowledges the influence of Van Helmont on his speculations of the fire; for instance, see [35].

raised some questions in regards to Van Helmont's seminal principles [31].[4] Generally, Helmontians viewed the book as being a development of Van Helmont's ideas and used them in conjunction [43].

3.4   An Impartial Reader's Assessment

A person of relative chemical knowledge living in Boyle's time would have recognised the tenor of many of Boyle's arguments as being primarily rooted in a Helmontian framework. This would immediately raise the question of bias, and imply that Boyle is basing his statements on the authority of Van Helmont. Indeed, it is interesting to see how Boyle supports Van Helmont's image as a chemical authority. Carneades' description of Van Helmont tends to be flattering; he describes him as 'one of the greatest Spagyrists that they [Chymists] can boast of' [44]. He often refers to the Flemish alchemist's recipes as 'great' [45], his experiments as 'more considerable than many Learned men are pleas'd to think him' [46] and his writing as 'faithful' or witty [47]. Even Boyle, separately from his alter-ego Carneades, describes Van Helmont as 'Bold and Ingenious' [48].

The suspicion of bias would also surround some of Boyle's 'evidence', often drawn on Helmontian sources. For instance, Boyle's enumeration of bodies that do not contain all three principles, such as gold, mercury and sand is largely a repetition of Van Helmont and other lesser authorities, and raises the question of whether Boyle had carried them out himself [49].

Boyle makes a better case against the *tria prima* when he refers to more conclusive experiments that he has clearly carried out. For instance, at pages 192–196 he talks about distilling box-wood and mixing the rectified liquor with powdered coral. From this experiment, he says, he obtained four active substances rather than three.

In any case, and to his credit, Boyle ultimately realised that his 'experimental proofs' against the *tria prima* were not very strong. Hence he went back to philosophical points to strengthen his argument. Indeed, his more compelling arguments are often drawn on philosophical considerations rather than empirical matters. Such are, for instance, his reproach on Paracelsians that their descriptions are too enigmatical and inconsistent, that their definitions are in fact summaries, or that their sulphurs, mercuries and salts are too diverse in bodies to be called by that name [50]. Nevertheless, it is strange that Carneades brought these philosophical arguments forth while maintaining that 'I would at this Conference Examine only the Experiments of my Adversaries, not their Speculative Reasons' [51].

---

[4]Boyle qualifies this criticism as referring to 'some Helmontians' at [41]. He must have known that Van Helmont certainly did not say that all things come from seeds, as he was a supporter of the theory of 'spontaneous generation.' On the debates on spontaneous generation in Van Helmont's time, see Hirai [42].

A final issue underlying Boyle's argument is its problematic generalisation of *tria prima* supporters. Who, precisely, did he have in mind as the object of his disapproval? The profile of the 'vulgar Chymist' comes across as a caricature that does not seem to fit any known alchemist in particular. Clericuzio argues that the writing was directed to the Oxford physiologists, chiefly Thomas Willis, but the latter supported the five-principle theory of de Clave and Villon rather than the generic *tria prima* [52]. Although this theory is briefly criticised as well, it does not make the direct object of the *Sceptical Chymist* [53]. Instead, it is Jean Beguin that seems to be uniformly criticised in the book. Beguin, the author of a very popular alchemical textbook, but deceased at the time of the *Sceptical Chymist*, was used as a type of 'straw man'. By focussing on this famous writer, Boyle implicitly criticised all the authors of contemporary alchemical textbooks. In this attack, Boyle again emulates Van Helmont [54].

More generally, Boyle's ire seems to be directed at a chemical attitude that rejected philosophical speculation. Carneades clearly despises those chemists that 'Confusedly Apprehend' (e.g do not have clear, philosophical ideas [55, 56]) and are not 'Learned Men' [56]. He believes that 'there is a great Difference betwixt the being able to make Experiments and the being able to give a Philosophical Account of them' [57]. Hence he criticises the chemists' tendency of calling themselves 'Philosophers' while not being able to argue their points philosophically. Carneades denies that the *tria prima* could account for complex phenomena including life, magnetism, fluidity, generation, gravity or space [58].

3.5 Conclusions: Theory Choice and the Tide of New Ideas

The *Sceptical Chymist* offers many reasons for Boyle's rejection of the *tria prima*, but none of them seem absolutely conclusive. An unbiased reader might be led to perhaps question the Paracelsian theory, but would not be convinced that the theory is false. Certainly many experiments that Boyle refers to could be questioned in their turn, while his Helmontian predisposition hinders the objective assessment of the arguments. The best the book achieves is to question some ordinary assumptions of the *tria prima* supporters, but an unbiased reader would need much more persuasion before choosing another theory in its stead.

This, of course, begs the question of why the *Sceptical Chymist* has been considered as a successful attack on Paracelsianism.[5] In fact, historians point that its positive reception was very limited in its period [59] and that its fame was exacerbated post-factum by nineteenth and early twentieth-century Whiggish history of science. Lawrence Principe points out that the book's strong praise had more to do with the early historians of science's need for a *Principia* in chemistry, which would

---

[5]See Principe [1] for a summary of the glowing praise of the book by early historians of chemistry.

clearly separate the old from the new [1]. It also has to do, of course, with the fact that Boyle situated himself on the 'winning' side of the debate between the *tria prima* adherents and their opponents. In Boyle's age, Paracelsian theories were coming under questioning by alchemical practitioners like Van Helmont and de Clave. One could even speak of a tidal wave of new thought that was sweeping away the ideas of the previous generation.

A good example in this case would be Boyle himself. The *Sceptical Chymist* does not present compelling evidence for Boyle's theory choice. I was unable to detect a truly game-changing experiment that would have made Boyle reject the *tria prima* theory. This might be because he was actually 'brought up', so to speak, in a different alchemical school, that of Helmontianism.[6] From 1650 onwards, Boyle was enthralled by the framework of Helmontianism and followed its tenets. In this sense, Boyle's case seems to support Kuhn's view that subjective factors colour individual theory choice [60]. Boyle's rejection of the *tria prima* seems to have been conditioned by the fact that he came 'under the influence' of Helmontianism at a young age.[7]

References

1. Principe LM (1998) The aspiring adept: Robert Boyle and his alchemical quest. Princeton University Press, Princeton, pp 27–29
2. Hooykaas R (1935) Der Elementenlehre des Paracelsus. Janus, 39, pp 175–188
3. Bianchi ML (1994) The visible and the invisible. From alchemy to Paracelsus. In: Rattansi P, Clericuzio A (eds) Alchemy and chemistry in the 16th and 17th centuries. Kluwer, London, p 18
4. Paracelsus (2008) Opus paramirum. In: Weeks A (ed and trans) Paracelsus: Theophrastus Bombastus von Hohenheim, 1493–1541: essential theoretical writings. Brill, Leiden, p 305
5. Pagel W (1984) Smiling spleen: Paracelsianism in storm and stress. Karger, Basel, p 92
6. Bianchi ML (1994) The visible and the invisible. From alchemy to Paracelsus. In: Rattansi P, Clericuzio A (eds) Alchemy and chemistry in the 16th and 17th centuries. Kluwer, London, p 19
7. Debus AG (1977) The chemical philosophy: Paracelsian science and medicine in the sixteenth and seventeenth centuries. Science History Publications, New York, pp 82–83
8. Boyle R (1661) The sceptical chymist. Cadwell, London, p 26
9. Debus AG (1967) Fire analysis and the elements in the sixteenth and the seventeenth centuries. Annals Sci 23:127–147
10. Joly B (1997) La chimie contre Aristote—La distillation du bois et la doctrine de cinq elements au XVIIe siècle en France. In: Bougard M (ed) Alchemy, Chemistry and Pharmacy,

---

[6]As Newman and Principe have shown, Boyle encountered Van Helmont very early in his natural philosophical career and was only for about a year under the influence of a non-Helmontian, Benjamin Worsley, before encountering the very Helmontian George Starkey. See [28].

[7]Kuhn [61] particularly emphasised the importance of subjective factors in paradigm choice. He pointed out that 'paradigm choice can never be unequivocally settled by logic and experiment alone' [62]. His argument that young scientists find it easier to adopt new paradigms seems particularly congruent in Boyle's case [63].

Proceedings of the XXth international congress of history of science (Liege, 20–26 July 1997), Vol XVIII, Brepols, p 70

11. For instance, Bostocke R (1585) The difference between the Auncient Physicke and the latter Phisicke. Robert VValley, London, chapter 8

12. Kuhn T (1977) Objectivity, value judgment, and theory choice. In: The essential tension. University of Chicago Press, Chicago, pp 321–322

13. Metzger H (1923) Les doctrines chimiques en France du debut du XVIIe à la fin du XVIIIe siècle, Paris: P.U.F, p 43

14. Debus AG (1977) The chemical philosophy: Paracelsian science and medicine in the sixteenth and seventeenth centuries. Science History Publications, New York, pp 79–80

15. Beguin J (1610) Tyrocinium chemicum. Paris, pp 19–24

16. Croll O (1609) Admonitory preface. In: Pinnel H (ed) Philosophy reformed and improved in four profound tractates, London, 1657

17. Sennert D (1619) De chimicorum cum Aristotelicis et Galenicis consensu ac dissensu liber. Wittenberg

18. Boyle R (1661) The sceptical chymist. Cadwell, London, p 179

19. Du Chesne J (1605) The practise of chymicall and hermeticall physick. Thomas Creede, London, chapter 13

20. Boyle R (1661) The sceptical chymist. Cadwell, London, p 187

21. Webster C (1966) Water as ultimate principle of nature: the background to Boyle's sceptical chymist. Ambix 13:96–107

22. Pagel W (1984). Smiling spleen: Paracelsianism in storm and stress. Karger, Basel, pp 14, 30

23. Clericuzio A (2000) Elements, principles and corpuscles: a study of atomism and chemistry in the seventeenth century. Kluwer, London, p 49

24. Van Helmont, JB (1652) Tria prima chymicorum. In: Ortus medicinae, 2nd edition, Amsterdam, p 329

25. Van Helmont JB (1652) Tria prima chymicorum. In: Ortus medicinae, 2nd edition, Amsterdam, pp 330–331

26. Van Helmont JB (1652) Elementa. In: Ortus medicinae, 2nd edition, Amsterdam, p 43

27. Van Helmont JB (1652) Tria prima chymicorum. In: Ortus medicinae, 2nd edition, Amsterdam, p 326

28. Newman W, Principe L (2002) Alchemy tried in the fire, Starkey, Boyle, and the fate of Helmontian chymistry. University of Chicago Press, Chicago, pp 207–272

29. Clericuzio A (2009) Les débuts de la carriere de Boyle, l'iatrochimie helmontienne et le cercle de Hartlib. In M. Dennehy & C. Ramond, eds. La philosophie naturelle de Robert Boyle. Paris: Vrin, pp 65–67

30. Clericuzio A (1994) Carneades and the chemists: a study of the sceptical chymist and its impact on seventeenth-century chemistry. In M. Hunter, ed. Robert Boyle Reconsidered. Cambridge: Cambridge University Press, pp 78–79

31. Boyle R (1661) The sceptical chymist. Cadwell, London, pp 225, 380–381

32. Boyle R (1661) The sceptical chymist. Cadwell, London, p 374

33. Boyle R (1661) The sceptical chymist. Cadwell, London, pp 49, 197

34. Boyle R (1661) The sceptical chymist. Cadwell, London, p 428

35. Boyle R (1661) The sceptical chymist. Cadwell, London, pp 382–384

36. Boyle R (1661) The sceptical chymist. Cadwell, London, p 73

37. Boyle R (1661) The sceptical chymist. Cadwell, London, p 91

38. Boyle R (1661) The sceptical chymist. Cadwell, London, p 75

39. Debus AG (1970) Science and education in the seventeenth century: the Webster-Ward debate, London: MacDonald

40. Boyle R (1661) The sceptical chymist. Cadwell, London, p 394

41. Boyle R (1661) The sceptical chymist. Cadwell, London, p 225

42. Hirai H (2011) Renaissance debates on matter, life and the soul. Leiden: Brill, pp 123–171

43. Clericuzio A (1994) Carneades and the chemists: a study of the sceptical chymist and its impact on seventeenth-century chemistry. In M. Hunter, ed. Robert Boyle Reconsidered. Cambridge: Cambridge University Press, pp 85–86
44. Boyle R (1661) The sceptical chymist. Cadwell, London, p 78
45. Boyle R (1661) The sceptical chymist. Cadwell, London, p 355
46. Boyle R (1661) The sceptical chymist. Cadwell, London, p 112
47. Boyle R (1661) The sceptical chymist. Cadwell, London, pp 78, 224
48. Boyle R (1661) The sceptical chymist. Cadwell, London, p [8]
49. Boyle R (1661) The sceptical chymist. Cadwell, London, pp 65, 345
50. Boyle R (1661) The sceptical chymist. Cadwell, London, pp 266–286
51. Boyle R (1661) The sceptical chymist. Cadwell, London, p 295
52. Clericuzio A (1994) Carneades and the chemists: a study of the sceptical chymist and its impact on seventeenth-century chemistry. In M. Hunter, ed. Robert Boyle Reconsidered. Cambridge: Cambridge University Press, p 80
53. Boyle R (1661) The sceptical chymist. Cadwell, London, p 287
54. Van Helmont, JB (1652) Tria prima chymicorum. In: Ortus medicinae, 2nd edition, Amsterdam, p 419
55. Boyle R (1661) The sceptical chymist. Cadwell, London, p 203
56. Boyle R (1661) The sceptical chymist. Cadwell, London, p 206
57. Boyle R (1661) The sceptical chymist. Cadwell, London, p 208
58. Boyle R (1661) The sceptical chymist. Cadwell, London, pp 301–315
59. Clericuzio A (1994) Carneades and the chemists: a study of the sceptical chymist and its impact on seventeenth-century chemistry. In M. Hunter, ed. Robert Boyle Reconsidered. Cambridge: Cambridge University Press, pp 84–87
60. Kuhn T (1977) Objectivity, value judgment, and theory choice. In: The essential tension. University of Chicago Press, Chicago, pp 325, 338
61. Kuhn TS (2012) The structure of scientific revolutions, 4th edition. Chicago, IL: University of Chicago Press
62. Kuhn TS (2012) The structure of scientific revolutions, 4th edition. Chicago, IL: University of Chicago Press, p 95
63. Kuhn TS (2012) The structure of scientific revolutions, 4th edition. Chicago, IL: University of Chicago Press, pp 90, 150

Chapter 4
Choice or No Choice? Affinity and Theory Choice

Georgette Taylor

4.1 Introduction

Falling as it does between two scientific 'revolutions', much of 18th century chemistry is often seen as transitional, lacking an easy to grasp character of its own. Tales of Newtonian assimilation dominate studies of the first half of the century, while those of the latter half are concerned in the main with the great changes to come. These two paradigm shifts tend to be smeared like grease across the historical lens, resulting in a soft focus view of events that obscures and subordinates all detail to the overriding narrative. Hence the tendency to Newtonianise 18th century chemistry evident in the works of Thackray [1, 2], Cohen [3] and Crosland [4, 5], and a similar over-emphasis on a monolithic phlogiston theory in many of the more popular accounts of the chemical revolution.[1] More recently, however, the imbalance has begun to be addressed, and other theories and practices that contributed to the 18th century discipline have been exposed to the light. One such theory that has received some attention in recent years is the theory (or doctrine) of affinity. Tracking this theory from its apparent origin over the course of the century poses a number of interesting questions about philosophically inspired narratives of the history of science, as well as shedding light more specifically on theory choice and development in 18th century chemistry.

---

[1]Affinity is notably absent from Hankins [6] for reasons which are far from clear.

G. Taylor (✉)
Department of Science and Technology Studies, University College London,
Gower Street, London WC1E 6BT, UK
e-mail: georgette.taylor@gmail.com

© Springer International Publishing Switzerland 2016
E. Tobin and C. Ambrosio (eds.), *Theory Choice in the History of Chemical Practices*, SpringerBriefs in History of Chemistry,
DOI 10.1007/978-3-319-29893-1_4

4.2　The Advent of Affinity

In 1718 Geoffroy presented to the Académie Royale des Sciences his "Table des Differents Rapports Observés en Chimie entre Differentes Substances" together with an explanatory *Mémoire* [7]. His paper inspired the Secretary of the Académie, Bernard le Bovier de Fontenelle to comment that "une Table Chimique est par elle-même un spectacle agréable à l'Esprit" [8].

The fact that some pairs of substances were more inclined to combine than others was a fundamental chemical axiom of long standing. Geoffroy tabulated these relations, listing empirically based generalisations of this knowledge in a grid of sixteen columns (Fig. 4.1 below). Each column was headed by a particular substance, and below this were listed other substances in order of their tendency to unite with the top one, from the most eager at the top to the most reluctant at the bottom. Geoffroy's paper explained his table and indicated in a very general way how the table had been created. He suggested that [7]:

> Par cette Table, ceux qui commencent à apprendre la Chimie se formeront en peu de temps une juste idée du rapport que les differentes substances ont les unes avec les autres, & les Chimistes y trouveront une methode aisée pour découvrir, ce qui se passe dans plusieurs de leurs operations difficiles à démêler, & ce qui doit resulter des melanges qu'ils sont de differents corps mixtes.

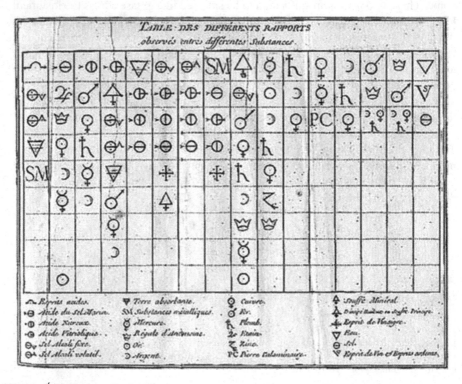

Fig. 1　Étiénne-François Geoffroy's 'Table des différents rapports' [7]

He did not at any point offer a causal explanation of how or why these relationships between substances operated. Instead, he proposed what he implied was a generalised 'loi' drawn from his observations [7]:

> Toutes les fois que deux substances qui ont quelque disposition à se joindre l'une avec l'autre, se trouvent unies ensemble; s'il en survient une troisiéme qui ait plus de rapport avec l'une des deux, elle s'y unit en faisant lâcher prise à l'autre.

According to the law, if two substances were combined and a third substance introduced that had more rapport with one of the combined substances than that substance had with its partner, it would oust the substance with less rapport from the combination and combine in its stead with the other. Knowledge of affinities could thus be used to manipulate matter, separating and combining different substances according to the chemist's desire.

It is notable that Geoffroy used the term 'rapport' to describe the tendency to combine together that exists between two different substances rather than either 'affinity' or 'attraction', both of which terms had well-recognised metaphysical baggage attached to them. The issue of language was of some importance to Geoffroy and his contemporaries, and as Duncan notes, Geoffroy went to some lengths to avoid any implied connection to any type of explanatory system [9, 10]. Presumably aware of the difficulties inherent in explaining these phenomena in ontological terms, he endeavoured to show that his generalised law could be useful even without drawing ontological conclusions. Avoiding the use of the more loaded terms of 'affinity' and 'attraction', instead, he referred only to 'rapports', a word that can be loosely translated as 'relationship' but which seems to have been intended to be synonymous with 'disposition à s'unir' [7].

It is interesting to note that in the *Proces Verbaux* of the Académie, where the paper was initially documented in manuscript, the change from the term 'affinité' to 'rapport' is recorded. Geoffroy began his reading of the paper on Saturday 27 August 1718, where the papers record that "M. Geoffroy a commencé à lirè un Ecris sur les differents degréz d'affinité des Matiéres Chimiquez" [11]. After a single blank sheet, the proceedings of the next meeting on Wednesday 31 August 1718 appear, and that "M. Geoffroy a achevé L'Ecris ... Suivant". The full text of the paper follows, in a different hand, titled as in the printed version, and referring to 'rapports' rather than 'affinité'.

In spite of Geoffroy's determinedly cautious terminology, his contemporaries did not follow his example [12, 13]. In 1723, the anonymously published *Nouveau Cours de Chymie* of Senac referred to Geoffroy's "Table des affinities des corps". Geoffroy's own term 'rapports' was, it seems, only ever used by him, being displaced by other, more etymologically controversial terms in almost every case. Perhaps inevitably, as well as 'affinity', 'attraction', or more commonly 'elective attraction' were often used, and by mid-century for many chemists the terms were effectively interchangeable. I shall continue to use the term 'affinity' as being sufficiently well associated with the theory that is the subject of this chapter as today 'attraction' is probably still burdened with too much physical (and metaphysical) baggage to be used with impunity.

4.3   Theory Choice and the 'New' Affinity

Geoffroy's paper was published only a year after it was read in the Mémoires of the Academie—remarkably quickly, given the oft noted sluggish speed of publication of this organ; we must assume that it was widely distributed and widely seen across scientific Europe. His Parisian colleagues seem to have afforded it a welcome reception—according to Senac, Geoffroy's table "a rendu plus de service à la Chymie qu'une infinite d'Auteurs par de volumes remplis de raisonnemens physiques" [13]. In terms of the 'adoption' of his table in Britain, however, the historical record is surprisingly silent. No letters seem to have passed, or debate to have been had amongst the luminaries of the Royal Society, for example, either acclaiming, decrying or even apparently taking notice of Geoffroy's paper. As Geoffroy was himself an FRS and corresponded with Hans Sloane until his death,[2] it might seem reasonable to expect to find some mention of his Mémoire in the papers of the Society. A number of his communications were published in the pages of the Philosophical Transaction, but there is no mention of his Mémoire therein. Neither here, nor in the Society's Letter or Journal Books can any reference be found.[3] The Society undoubtedly received copies of the Mémoires of their Parisian counterpart, and as the Journal Books show, they often read the papers out at their regular meetings.[4] Given Geoffroy's close ties with the Royal Society, it seems strange that his table was not afforded a welcoming reception in Britain on its publication; or indeed that it was not afforded any reception at all.

Why then, was Geoffroy's table so comprehensively ignored by the natural philosophers of Britain? Duncan has suggested that [14]:

> Geoffroy's table was presumably less well known in Britain ... and perhaps also the notion of affinity which was thought to be expressed in Geoffroy's table (though he does not use the word himself) was still felt in Britain to be in some way contrary to the notion that chemical combination was due to attraction between particles.

The problem for the historian arises in part from the fact that in Britain there was little public forum for discussion of non-mechanistic chemistry. This does not of course mean that there was no such art practiced, or indeed that its practitioners did not find Geoffroy's paper of interest. Unfortunately though, it does mean that there is little or no evidence on which the historian can draw to settle the matter. The silence may well be merely the silence of history and it is perhaps rather superficial to insist that it reflects the silence of the 18th century actors.

There is another question to be considered, however; how 'new' were Geoffroy's ideas? Geoffroy did not dwell in detail on theory, and the table was described

---

[2]The Royal Society holds some of their correspondence, as do the British Library as part of the Sloane Collection.

[3]My researches into both the published and unpublished papers of the Royal Society have shown that although the Society did receive a letter from Geoffroy to Sloane in 1719 (Royal Society of London 1718–1721), no discussion of Geoffroy's paper or comment was noted.

[4]See e.g. December 21st 1721, Royal Society of London 1718–1721.

simply as a synopsis of chemical phenomena. These phenomena were familiar to all those who practised chemistry. Both Robert Boyle and Isaac Newton had demonstrated their acquaintance with series of precipitations and solutions; most famously in Newton's Query 31 appended to his 1706 edition of Opticks. The lengthy Query 31 contains Newton's speculations on matter and its attractions. About midway, he turned to chemistry, suggesting that inter-particulate attractions might be responsible for the successive precipitations of metals from solution in acid [15]:

> And so when a Solution of Iron in *Aqua Fortis* dissolves the *Lapis Calaminaris*, and lets go the Iron, or a Solution of Copper dissolves Iron immersed in it and lets go the Copper, or a Solution of Silver dissolves Copper and lets go the Silver, or a Solution of Mercury in *Aqua fortis* being poured upon Iron, Copper, Tin, or Lead, dissolves the Metal and lets go the Mercury; does not this argue that the acid Particles of the *Aqua fortis* are attracted more strongly by the *Lapis Calaminaris* than by Iron, and more strongly by Iron than by copper, and more strongly by Copper than by Silver, and more strongly by Iron, Copper, Tin and Lead, than by Mercury?

Many historians have seen Geoffroy's paper as stemming from the Newtonian natural philosophy sweeping Europe. That Newton's notion of attractive force influenced many later chemists' affinity theories is undeniable; affinity lent itself too easily to an ontology of particles and attractive forces for it to be otherwise. Three strands can be discerned in the historiography that ties Newton to affinity theory. The first asserts that the idea of ordering affinities originated with Newton, and that Geoffroy merely rearranged Newton's words and ideas in tabular form. The second, in most cases deriving from the first, assumes that Geoffroy himself was a 'Newtonian'. A third assumption that any espousal of an affinity theory necessarily involved a commitment to a Newtonian ontology is also common.

It is beyond the scope of this chapter to demolish these particular assumptions in detail, but the first is perhaps worthy of comment.[5] As Guerlac and Klein have shown, there were many potential sources of inspiration for Geoffroy's table besides Newton's Opticks [17]:

> Much, if not most, of the information in the table he could have drawn from the seventeenth-century chemical tradition, as indeed Newton himself had done in accumulating the chemical facts that he set forth in Query 31.

Indeed, a relatively cursory glance at the writings of Robert Boyle will discover similar observations, for example his "Of the Mechanical Causes of Chemical Precipitation" [18] set out a detailed mechanical explanation of a model of preferential combination that would have been recognized by later chemists as an affinity separation [19]:

> another way, whereby the dissolving particles of a menstruum may be rendered unfit to sustain the dissolved body, is to present them another, that they can more easily work on.

---

[5]For a comprehensive discussion (and demolition) of these assumptions, see my Ph.D. dissertation [16].

He continued by giving an example of how the recognition of the relationships between the dissolved body, the menstruum and the precipitant was of practical importance to the practitioners of metallurgy [19]:

> that in these operations, the saline particles may really quit the dissolved body, and work upon the precipitant, may appear by the lately mentioned practice of refiners, where the aqua fortis, that forsakes the particles of the silver, falls a working upon the copper-plates employed about the precipitation, and dissolves so much of them, as to acquire the greenish blue colour of a good solution of that metal. And the copper we can easily again, without salts, obtain by precipitation out of that liquor with iron, and that too, remaining dissolved in its place, we can precipitate with the tasteless powder of another mineral.

So, in the face of this, why do historians insist on the Newtonian origin of affinity theory? There is a recognised tendency for ideas to be deliberately linked to revered authorities in order to acquire prestige or validation [20]. I would suggest that the 'Newtonianisation' of Geoffroy and his table is such a case. The third strand mentioned above seeks to extend this classification to all affinity theories. It is unfair, however, to condemn modern historiography for the widespread assertion that Newton 'invented' affinity. Many 18th century chemists asserted something remarkably similar. John Warltire, a public lecturer in chemistry, stated unambiguously that [21]:

> The Plan of this Table was first given to the World by the illustrious Sir *Isaac Newton*, in his Optics, Quere 31st; and has received many Improvements from *Stahl, Geofroy*, the Edinburgh Chemists, and others.

and many British chemists in particular took care to assert Newton's authority over the origin of affinity theories.[6] It is tempting to attribute this trend to nationalistic feelings, to suggest that British chemists wanted to claim such a useful chemical theory for their own, and it is certainly the case that Rouelle, who taught affinity in France from the 1740s was avowedly not a Newtonian [23]. Whatever the cause, it is clear that there is a strong tradition of ascribing the origin of affinity to Newton on the strength of the 31st Query. As those temporally closer to Newton than we are today instituted this myth, so some modern historians have accepted their assessment without demur, and the fiction has been propagated.

Whatever the reason for the lack of comment on its first publication, it is undoubtedly the case that Geoffroy's affinity table took some time to take root in 18th century chemistry, whether British or French. Why is this interesting? Because any examination of any chemical textbook from the middle of the century onwards will demonstrate that affinity by this time was deemed to be at the heart of the discipline. In the latter half of the 18th century, and beyond, a chemistry book without an affinity table, or indeed many affinity tables, was a novelty. As Kim says, "affinity tables... became a fashion" [24]. Affinity tables were omnipresent. From being largely ignored on publication, affinity tables (and their concomitant assumptions, both necessary and supplementary, which we might describe as the

---

[6]See, for example, Cullen [22] in which Cullen asserts the role of Newton in the formulation of affinity.

theory of affinity) became fundamental to the discipline, part of its history and regarded by all chemists as essential to its future. So, how did affinity become the all-conquering monster it undoubtedly became?

At first sight, perhaps Geoffroy's table seemed little different from Boyle and Newton's lists of displacements. While it was historically innovative in that it was the first of its kind, it may not have been immediately seen as particularly novel. Appearing in a 'research' context, it appeared out of place, neither fish nor flesh. Presented as an aide memoire, perhaps it seemed just too familiar to be significant and perhaps this very familiarity bred contempt. Even in France, as Fontenelle extolled the utility of "une Table Chimique", he looked past Geoffroy's table, anticipating "une Table de Nombres ordonnés suivant certain rapports ou certaines propriétés" [8]. Geoffroy's table can only retrospectively be seen as pioneering: to its contemporaries it was simply an articulation and ordering of knowledge that had been common for some time. However, Geoffroy and his contemporaries did appreciate that it might be useful to novices.

Although Geoffroy was Professor of Chemistry at the Jardin du Roi, it was not in this guise that he presented his table. His *Treatise* refers neither to the table nor to affinities or rapports [25]. Nevertheless, he called it "une chose fort utile", and carefully distinguished between its use for "ceux qui commencent à apprendre la Chimie" (i.e. its pedagogical function) and its use for "les Chimistes" [7]. The table was to provide novices with a general survey of the relationships between substances [25]. For mature chemists it would enable them to discover the "mouvements cachés" when substances are mixed, and to predict the results of their mixtures.

Geoffroy's comments about his table's pedagogical utility proved to be prescient.[7] The earliest reference to it in British texts were in a pedagogical context and those who sought to use it in that context were responsible for drawing out the novelty in their later emphasis on the way the tables and the law were to be used. Similarly, in France, what might be termed the 'revival' of the affinity table, some decades after Geoffroy's original publication, was in the lecture halls and (later) in textbooks.

Although by no means commonplace, in the early decades of the 18th century, chemistry in Britain was (happily) not confined to the activities of the Royal Society or the official (lamentable, for the most part) teaching in the universities. There were of course many working practitioners, whether apothecaries, assayers, physicians or any of the other skilled occupations that involved the manipulation of matter. These, in truth, were the artisans who might be more likely to describe themselves as 'chemists'. In spite of the lack of a formalized, specifically chemical 'establishment', there was also a thriving unofficial network of chemistry lecturers.[8] Many of its luminaries were self-educated while others had learned chemistry as part of their medical training. They contributed to the dissemination of chemistry

---

[7]For detail, see Taylor [26].
[8]For a general survey of 'itinerant' lecturers, see Gibbs [27].

through the performance of public lecture courses, and published their lectures and course syllabi.[9] By the 1730s, in spite of the lack of any 'establishment' encouragement, Geoffroy's affinity table had begun to creep into this informal pedagogy. Its first mention in a British publication was by the most prominent (and successful) lecturer in chemistry of the first half of the 18th century, Peter Shaw.

Shaw's career illustrates the contrast between the social situation of the chemist within (or more accurately, without) the English scientific world when compared to France. He promoted himself as an indispensable intermediary between the proponents of chemical knowledge, and the society physicians who were his patrons [33]. As Golinski has noted, Shaw's choices of works for translation and publication did not emphasise a predominantly Newtonian stance. Instead, they took for the most part what might be termed a 'chemical' position that followed the Baconian inductive method and emphasised a qualitative chemistry, often founded on systems of elements [33]. His entrepreneurial role served to bring him to the attention of the establishment, and enabled him to achieve similar social heights, at which point his interest in chemistry apparently had served its purpose and he published no more.

While still engaged in his role as public lecturer in chemistry, Shaw published a selection of headings for a proposed course in "philosophical chemistry". Amongst the headings for this presumably theoretically biased course, he included [34]:

III A View of the different RELATIONS, vulgarly call'd Sympathies and Antipathies, or Attractions and Repulses, observ'd betwixt different chemical Bodies; with the uses of this Doctrine in Philosophy and Chemistry. See Boyle, Hook, Homberg, Newton, Stahl, and the Memoir of Geoffroy in the Works of the Royal Academy for the Year 1718.

Whether a course was ever given along these lines we do not know. Shaw did indeed give courses on chemistry for many years in London and later in Scarborough, and he published the syllabus to one such course in 1733. Under the heading 'synthesis' he stated [35]:

it is proper to enquire what other Bodies there are which may be perfectly separated into different Parts by the way of Menstruum, Absorbent, or Precipitant; so as to leave the separated Matters unalter'd in their Natures; and fit to compose the original substance again. This Enquiry depends upon finding out the secret Relations which exist betwixt particular Bodies; and these Relations can only be discover'd by particular Trials.* [*See M. Geoffroy's Paper to this purpose in the French Memoires].

These are the first references to Geoffroy's *Mémoire* in any British publication, some fifteen years after its initial presentation. It cropped up again in Shaw's 1741 translation of Boerhaave's *New Method of Chemistry*.[10] Although Boerhaave's work advised his readers to consult Geoffroy's works, it did not comment

---

[9]For example, see [28–32].

[10]Shaw's unofficial translation of the lectures [36] provided equally copious notes including references to some of the Memoires, but the latest reference is 1716.

specifically on the 1718 *Mémoire*.[11] However, Shaw's notes (often vastly out of proportion to the Boerhaavian part of the work—some pages allow Boerhaave the top line, while Shaw's notes and updates occupy the remainder of the leaf) directed the reader to many useful chemical papers including Geoffroy's 1718 paper. He describes the table here as [38]:

> a system or table of the mutual relations betwixt different substances in chemistry; which, if rightly understood, and carried on, might become a fundamental law for chemical operations, and guide the operator with success.

It is clear that by the 1730s and into the 1740s Shaw, at least, was recommending Geoffroy's affinity table. He apparently viewed the table as a useful synoptic tool, albeit of potentially heuristic value, suggesting that it had the potential to form the basis of a "fundamental law" which could "guide the operator". The notion of discovering from the patterns of relations a law that could predict chemical behaviour must have been an attractive one, although Shaw apparently made no effort himself to deduce such a law and although he referred his audiences to Geoffroy's table, he did not reprint it in any of his various publications.

Shaw was followed quickly by other lecturers, both high and low in status, formal and informal in position. William Lewis, a highly regarded apothecary based in London who had been offering public lectures in chemistry in the 1730s and 1740s and was often consulted in chemical matters by the Society for the Encouragement of Arts, Manufactures and Commerce was (probably) the first to republish Geoffroy's table as part of a textbook, in his *New Dispensatory* of 1753. Lewis became aware of Geoffroy's table prior to 1748, when he published his proposals for printing his *Commercium Philosophico-Technicum*. This was an ambitious project, a periodical publication to be published in six parts per annum, and "designed as an attempt to advance useful knowledge" [39].[12] Lewis's vision was of an empirical, pragmatically utilitarian chemistry, and the synoptic tabulation of knowledge formed an intrinsic part of his design. The *Proposals* specifically set out his intention to include in the work a [41]:

> table of the relations or affinity which different substances bear to one another; with an experimental account of its uses.

It seems likely that affinity was included in Lewis's public lectures, but the historical record is silent on this. Sadly, only the first few parts of *Commercium Philosophico-Technicum* were published in 1763 and, in spite of the fact that an extensive explanation of affinity appears in the preface, no table was included.[13] However, as part of his report on the properties of platina that occupied almost half

---

[11]See John Powers' recent work on Boerhaave's chemistry in which he provides an explanation of Boerhaave's apparent lack of interest in affinity [37].

[12]When the work finally appeared, this was amended to read "an attempt to improve Arts, Trades and Manufactures" [40].

[13]The discussion of affinity appears as part of Lewis's explanation of the different types of "active powers" of bodies, chemical and mechanical [42].

of the work, Lewis inserted fifteen separate columns, each of three substances, showing the affinities of this curious metal alongside explanations of how he ascertained them.

In the first edition of the *New Dispensatory*, affinity is mentioned only in the introduction with specific reference to the reactions of marine acid. Lewis commented that [43]:

> the doctrine of the affinity of bodies is of very extensive use in the chemical pharmacy: many of the officinal processes are founded on it[14]

Lewis slightly changed the form and content of Geoffroy's table, using text rather than symbols, and setting out the orders of affinity in rows rather than columns. Sivin has carefully compared Lewis's table with Geoffroy's, highlighting those columns and substances that Lewis removed or added [45]. Sivin's analysis is unfortunately marred by his somewhat present-centred reference to affinity table columns as "displacement series". Thus he deplores Geoffroy's failure to recognise the effects of mass action, and rather oddly accuses both him and Lewis of inconsistency in the columns for, respectively, iron and antimony, and nitrous and marine acids.[15] Only Sivin's misinterpretation of the columns as "displacement series" can explain this mistake. It would appear that the theoretical assumptions required to enable the construal of affinity tables are counter-intuitive to those more versed in the chemistry of later centuries.

In later editions of the *New Dispensatory*, Lewis made further amendments to his table, which by this time bore little resemblance to Geoffroy's. He also extended his discussion of chemical theory, and the table appeared within a newly introduced 'Elements of Pharmacy', which constituted about a third of the total work. The table, now in the form of 19 textual lists, was included under a separate entry 'affinity' with a brief but clear explanation of how the table (but notably not affinity itself) worked. Lewis shied away from causal explanation, remarking only that [46]:

> the power in bodies, on which these various transpositions and combinations depend, is called by the chemists affinity; a term, like the Newtonian attraction, designed to express, not the cause, but the effect

Although he referred to 'Newtonian attraction' in the same breath as affinity, it is clear that he was not conflating the two. The connection between the terms was in their reference to effect rather than cause. This is the same pragmatic approach to

---

[14]The word 'officinal' is defined by the OED [44] as referring to a medicinal preparation. "kept as a stock preparation by apothecaries or pharmacists (now *rare*); made to a standard prescribed in a pharmacopoeia or formulary, included in a pharmacopoeia." "officinal, a."

[15]Geoffroy's table included a column headed with iron, with regulus of antimony immediately below it, and copper, silver and lead below that. The next column was headed by regulus of antimony, with iron immediately below it, and again copper, silver and lead below that. There is nothing inconsistent about this. Geoffroy was clearly showing that the strongest affinity lay between antimony and iron, and that this will lead to all other metals being removed from their union with either of these two metals. Indeed, there would have been something wrong with Geoffroy's table if these two columns had not borne this relationship to each other.

causality that seems to have characterised chemistry of the 18th century, and it might well be argued that the discipline was all the better for it.

Tracing the 'revival' (the term seems justified, as it seems to have occurred decades after the original publication, and indeed some years after Geoffroy's death) of affinity backwards, then, leads us into the pedagogical realm. In France and Britain by the 18th century, there was an appreciation that chemistry had to be taught: apothecaries provided an important service to the medical profession, and physicians also had an need for some familiarity with the substances they prescribed. In Scotland, men of industry, prompted by new legislation on manufactures, were beginning to discover a need for greater understanding of chemistry.[16] Entrepreneurial spirits abounded in Britain at this time; private lecturers like Shaw and Lewis (and their contemporaries in France) perceived the great potential utility of chemistry, and both turned it to their own advantage. In their eyes, a synoptic table that summarised so much practical chemical knowledge could not fail to be useful. Thus it fell to the pedagogues, the presenters of lectures and the writers of textbooks to introduce Geoffroy's table to chemists.

It has been noted that the periodic law and table first appeared in Mendeleev's textbook *Principles of Chemistry* and it has been speculated that it was the specific pedagogical demands of preparing a "general survey of chemistry" that prompted his discovery of the law [48]. Bensaude-Vincent has argued that Mendeleev's position as a teacher confronted him with questions unlikely to face other chemists: "How to summarize chemistry? How to order the chapters?" [48]. Chemists of the 18th century, before the rationalisation of chemical nomenclature, were in an even more confusing environment. Those who began to teach the science had to formulate answers to similar questions, and affinity helped to provide these answers.

In Britain, the most influential teacher of chemistry from mid-century onwards was William Cullen. Cullen embedded affinity at the heart of his lecture course, treating it as entirely fundamental to the discipline. Cullen, like Lewis, extended the practical range of Geoffroy's table, producing new columns setting out the affinities of substances not previously included. But where Lewis conceived of affinity as a utilitarian ordering or classification of empirically based facts, Cullen developed a complex theory, linking it explicitly to his observations of chemical practice. He extended its theoretical range by the inclusion of new components covering such matters as complex affinities and the role of heat and it moulded and dictated much of his didactic strategy.

Cullen taught that the chemist investigated the "particular properties of bodies" [49] and the structuring role that he allocated to this theory is clear from his carefully worded definition [50]:

---

[16]Two Acts of 1727, the first for the better Regulation of the Linen and Hempen Manufactures in Scotland, and the second for Encouraging and promoting Fisheries and other Manufactures and Improvements in Scotland were crucial here. The latter Act resulted in the creation of the Board of Trustees for Manufactures, a body which paid both Cullen and Black premiums for investigations into bleaching. See [47].

the changes of the Qualities of Bodies, produced by Chy, are all of them produced by, Combination, or Separation. The Office of Chy is to induce new qualities on bodies, & take away old ones; & this, I say, it does by Combination & Separation.

Cullen also classified the specific operations and processes in terms of his affinity theory. In both analysis and synthesis, the primary tool of the chemist was the power apparently inherent in substances to combine together, but fire had a place in affinity theory too, promoting fusion or solution to enable the affinities to act [51]:

Combination depends upon Attraction, and this upon Fluidity, wch is employ'd in Solution or Fusion

Separation depends upon Elective Attraction or the Action of Fire.

His pedagogical endeavours effectively retrieved affinity from obscurity and instituted a new conceptualisation in which the two uses suggested by Geoffroy coalesced.[17] The doctrine of affinity provided him with a didactic framework, structuring his teaching of the discipine as well as consolidating and assimilating innovation. He used affinity to organise and provide structure to his course, supplying his students with a large (sometimes pre-printed) affinity table and he referred throughout his course to the actions of affinity.

Cullen began teaching at Glasgow University in 1748, thirty years after the original publication of Geoffroy's paper. In 1749 Joseph Pierre Macquer published his *Elemens de Chymie Theorique* in France which included not only a reproduction of Geoffroy's table, but also an entire chapter offering instructions for its use. Duncan has suggested that "the efficient cause of the revival of interest in affinity in the 1750s was probably the publication of Macquer's Elemens" [52]. The work was undoubtedly influential; one of the few works that Cullen recommended to students, although with stern criticism of some aspects [53]. Macquer's chemistry was notable for its consistency, making use of affinity theory throughout. He asserted that [54]:

Nous avons vu dans le cours de cet Ouvrage, que Presque tous les phénoménes qu présente la Chymie, sont fondés sur les affinities qu'ont ensemble les différentes substances, sur-tout celles qui sont les plus simples.

He added to Geoffroy's table, a set of six (seven in the English translation) rules or propositions relating to the action of affinity "quelle qu'en soit la cause". He referred to these propositions as "vérités fondamentales" [55], according them an unusual value that perhaps proceeded from the fact that they allowed him to link his affinity to his phlogiston theory as well as to his Aristotelian matter theory. Macquer took a holistic view of his chemistry, and his affinity theory provided a crucial link between the observations of matter as it tended to be found in the laboratory: messy, complicated, and a long way from the perfectly pure ideal of the Aristotelian elements; and the a priori ontology that he believed lay behind these untidy observations.

---

[17]For detail on Cullen's use of affinity theory in his pedagogy, see Taylor [16, 26].

Macquer's reliance on affinity theory undoubtedly originated in his own training, which included attendance at lectures by Guillaume Francois Rouelle. From the 1740s Rouelle's lectures at the Jardin du Roi had, like Cullen's, been imbued with affinity; his slightly revised version of Geoffroy's table was published in the *Encyclopedie* [56]. Rouelle is considered to have been enormously influential in France primarily through his teaching; as well as Macquer, he is known to have taught a number of the luminaries of late 18th century French chemistry including both Lavoisier and Richard Kirwan, two sides of the later revolutionary debate [57]. Rouelle might legitimately be seen as Cullen's counterpart in France.

However, for Cullen and his successors, the table and the law provided in Geoffroy's *Mémoire* were insufficient for affinity theories to be useful. From Macquer and Cullen onwards, lecturers and writers of textbooks added extra components to their theories, articulating tacit assumptions, and generally fitting affinity for a useful purpose. Macquer's 'fundamental truths' are an example of these kinds of developments. Cullen too offered further rules about the operation of affinity, and extended its use beyond mixtures of three substances to explain and predict the behaviour of two compound (or up to four separate) substances when mixed together. He developed new diagrams and 'equations' that offered students a shorthand way to depict operations, as well as a heuristic paper tool to predict and assist in the interpretation of complex mixtures and combinations.[18]

As pedagogical tools, affinity theories dictated the structure of courses, the explanations of observations, and guided the students in their operations [16, 26]. Affinity tables were used almost as instruments, directing chemists in their search for novelty and in the interpretation of experimental results. As such, affinity crept out of the classroom and into the research laboratory, from the pages of textbooks and (although the demarcation between the two was far from clear at this time) into what we might (albeit somewhat anachronistically) term research publications. New substances were investigated, following William Lewis's early example, with a view to ascertaining their affinities, and substances were only regarded as 'chemical' once they had been characterised and could henceforth be identified on the basis of their affinities. At the same time, affinity was itself an object of investigation, as new tables were produced, and new components were formulated to account for new observations. Newly discovered phenomena were explored and assimilated into affinity theories as the century wore on.

As an example, on the one occasion where Cullen (for reasons unknown) apparently digressed and departed from his customary avoidance of speculation in his lectures, he set forward a highly unusual hypothesis on the nature of heat and phlogiston. His theory was complex but endeavoured to explain the various physical phenomena of heat as well as chemically generated heat and inflammation in material terms. He concluded, with the help of his affinity theory, that the so-called phlogiston (on which subject he was always doubtfully dampening in his

---

[18]On Cullen's diagrams, see [4]. On paper tools in chemistry, see [58–60].

lectures) was no more than a combination between two ordinary substances, an acid and the newly discovered 'mephitic' or fixed air. This notion explained the difficulty chemists apparently had in isolating phlogiston, as mephitic air did not unite with acids in isolation. However, Cullen (and others) claimed that where two substances could not be made to combine together, when one was combined with a third, which itself had some affinity for the remaining substance, this would facilitate the combination. So, while Cullen's phlogiston could never be isolated, it could always exist in combination with another substance.[19] Cullen thus used his affinity theory in two ways: to guide and direct his pedagogy, and his students' learning; and to similarly guide and direct his own speculative experiments and theorising.

By this time, affinity was used routinely by chemists in their judgements and interpretations of observations made outside the lecture hall. A particularly nice example appears in some correspondence between Joseph Black and James Ferguson in which Black explains that while corrosive calcarious earths will violently attract and dissolve in all acids, when spar or marble (mild calcarious earths, that is calcarious earths in combination with mephitic or fixed air) are placed in distilled vinegar, they will dissolve eventually, but it takes a very long time because:

> there is no substance that attracts & retains mephitic air so strongly as do the calcarious earths, they therefore attract it very near as strongly as they do those very weak acids (by weak I do not here mean dilute but less active or powerful) & hence these acids expel the air slowly & as it were with difficulty[20]

Thus individual substances, and whole classes of substance were characterised on the basis of their comparative affinities and surprising or unusual observations were explained by reference to the complex operations of affinities in competition.

4.4 Beyond Pedagogy

The assimilation of the doctrine of affinity into chemistry was thus initially prompted by pedagogical needs and its role thereafter accelerated and expanded both within the pedagogical context and beyond. This clearly does not sit well with the Kuhnian view of pedagogy as passive reflector of scientific consensus, contributing little or nothing to the progress of the science. With regard to affinity, pedagogy and pedagogical needs served to drive forward the discipline, in contrast to Kuhn's argument that textbooks epitomise 'normal science' and that revolutions occur outside the classroom.

Kuhn's division of science into normal and revolutionary varieties has perhaps served to concentrate historians' minds on the 'revolutions' to the detriment of the

---

[19]For a detailed account of Cullen's speculative theory, see Taylor [16].

[20]Black to Dr. James Ferguson, Glasgow 14 October 1763 [61].

'normal science'. The doctrine of affinity was fundamental to 18th century chemistry, and remained impervious to the more 'revolutionary' developments that ordinarily receive top billing in the received view of 18th century chemistry. Neither new airs nor new nomenclature were to depose the doctrine of affinity, which remained omnipresent well into the 19th century. Indeed, for Thomas Thomson, writing in 1830, affinity still constituted "confessedly the basis of the science" in spite of the fact that "it had been almost completely overlooked by Lavoisier" [62]. It is clear that affinity theories provided sufficient continuity to the discipline to enable avoidance of the complete disjunction that seems to be implied by Kuhnian historiography.

On the other hand, the doctrine of affinity itself was riddled with discontinuity and divergence. Each chemist's understanding of affinity differed from the other in certain details, and while this might seem to be a flaw or weakness in the theory, I would argue that the very flexibility of affinity was perhaps the key to its continued success. We have seen that Geoffroy offered a 'law' of affinity in his 1718 *Mémoire* [63]:

> Every time that two substances which have some disposition to join with each other happen to be united together, if there supervenes a third which has more relationship [*rapport*] with one of the others, it unites with it and makes it release its hold on the other.

However, there were other assumptions that needed to be made before Geoffroy's ideas could be adopted and applied usefully. Certain concepts were required to be 'filled-in' for affinity theories to be operationally intelligible. Many of these were tacit assumptions that had been made by Geoffroy himself in putting his original paper together. For example, Klein argues that Geoffroy's affinity table articulated a new conception of "chemical combination, compound, and reaction" [64]. In this scheme, the chemical compound was made up of empirically "homogeneous chemical substances", i.e. relatively stable substances which could be combined to create new substances, and recovered from these new substances without alteration [65]. We have seen that Cullen and Macquer supplemented Geoffroy's law with their own assumptions, rules of thumb, and 'fundamental truths' to answer questions like: what counts as a compound or a mixture? How is the chemist to know the difference? What happens when more than three substances are mixed? As substances are ordered in affinity tables, how is this order established? And on what empirical basis?

Those who intended to utilise affinity theory in their chemistry often formulated extra theoretical components to answer these questions. Macquer's rules, and Cullen's extra law about the use of a third substance to persuade otherwise antipathetic substances to combine are examples of these kinds of assumptions. However, not all these additional components were necessarily articulated in their publications, or even in their lectures. Many were tacitly assumed and can only be teased out of any particular chemist's theory by a close analysis of their accounts of their practice and by careful exegesis of their interpretation of observations. Thus affinity tables became enlarged, refined and amended according to each chemist's own bundle of tacit assumptions. Occasionally they were supplemented with entirely new theoretical components in response to new empirical and theoretical information. By this time, consensus assumed the doctrine of affinity to be foundational to chemistry; in particular consequence of the pedagogical context in which affinity theories became omnipresent, explanations of new phenomena were set forth in terms of the doctrine of affinity.[21] Affinity theories thus began to reflect and reinforce the invisible boundaries around the discipline. Examples of these 'supplementary' assumptions can be found by considering the role of heat in 18th century chemistry.

Geoffroy had not included any component dealing with heat in his 1718 affinity theory. Nevertheless, it had long been appreciated that heat affected chemical combination; indeed, the traditional agent of the chemist was fire. Nicknames like 'philosophers by fire', 'sooty empirics', even 'puffers' [67], emphasised chemists' near-permanent station beside the furnace.[22] Operations like distillation and sublimation had been used for centuries to 'resolve' complex substances into their principles or constituents. In his efforts to separate and extract active 'essences' from complex natural bodies, fire, the "grand agent in the resolution of bodies" [69], was the chemist's friend.

William Cullen taught his students that chemistry relied on the 'agencies' of affinity and fire, and the continuing importance of heat to the practice of chemistry in the 18th century is evident from the variety of furnaces available. Lewis listed five common types (the open fire, sand furnace, melting furnace, still, and athanor) [70] each designed to produce a particular range of temperatures. Cullen devoted four of his 1766 lectures to different types of furnace [71] and Black developed one to his own design.[23] The main method of controlling the intensity of heat was the choice of furnace employed, although this was gradually refined by technological improvements and theoretical advances.

Traditionally, heat could 'raise' substances, separating the volatile from the fixed, as in distillation and sublimation. This physical effect of expansion was well understood and many chemists tied this into their affinity theory. As we have seen, Cullen insisted that for combination to take place one or both substances must be fluid, and some variant of this claim was included as a component of most affinity theories. This necessary fluidity was achieved by solution or fusion; the former often, and the latter always, requiring the application of heat. Thus recalcitrant substances that were reluctant to combine when cold could be induced to unite by the addition of heat. But by the 1760s it was becoming widely acknowledged that the presence or absence of heat also disrupted the way in which affinities acted. In chemical operations the addition (or removal) of heat seemed to change the affinities of bodies, and this in turn modified the outcome of combinations. This fact is nicely demonstrated by Torbern Berman's magnificent affinity table which

---

[21]On the crucial relation between training and scientific practice, see Galison and Assmus [66].

[22]A recent work covering the 'history of chemistry from alchemy to the atomic age' was titled "Creations of Fire", presumably in homage to the earliest methods of chemistry [68].

[23]One of Black's furnaces cost Joseph Priestley £3 13s 6d [72].

showed affinities in the 'wet way' and the 'dry way' (i.e. in solution, or in fusion) separately [73]. The effects of heat on mixtures were predictable only through long experience, and subject to no known universal law. As Nicholson noted [74],

> the variations of temperature, ... tend greatly to disturb the effects of elective attraction. These causes render it difficult to point out an example of simple elective attraction, which may in strictness be reckoned as such.

Although these effects were bewildering, chemists seem to have been generally sanguine about the possibilities of rationalising them. Indeed Nicholson airily predicted that [75]:

> doubtless, by separating their parts, it will not be difficult to explain the effect of heat upon the change produced in bodies by their elective attractions.

In spite of his optimism, Nicholson seems to have eschewed any attempt to systematise the chemical effects of heat. There were, however, a few venturesome philosophers who did attempt to generalise and clarify these phenomena. They tried to preserve the status of their affinity theories as useful tools by adding new components that might enable them to predict the results of the conjunctions of heat and affinities.

Even if heat was not applied by the chemist, certain operations were apparently capable of producing perceptible heat or cold, e.g. the slaking of quicklime in water, which produced a great deal of heat.[24] Newton had speculated that this phenomenon was due to the speed with which the particles of each substance approached each other in consequence of their mutual attraction [77]. With Black's work on specific and latent heats[25] a material view of heat became predominant in natural philosophy [82, 83]. Where such a material heat was postulated, it was often considered as a constituent of chemical combinations, capable of expulsion like any other substance. Black theory was often seen as endorsing the material ontology as heat capacities could be seen as the tendencies of heat to combine with ordinary matter. Perhaps heat's own affinities were responsible for its inclusion or expulsion from combinations? Some chemists began to include the affinities of this material heat in their calculations, turning what had previously been thought of as simple combinations and separations into more complex systems.

The effects of heat needed to be examined, rationalised and codified, to be brought under control as affinities had been. Then chemists would be truly in control of matter. Rules were formulated attempting to codify the various roles of heat as part of chemical practice. Many of these rules were specifically designed for inclusion in affinity theories. The weight of the anomalies mounting between theories of heat and affinity prompted chemists to place an even greater reliance on the

---

[24]This particular instance of the generation of heat was a common interest for chemists of both the 17th and 18th centuries, discussed and explained in various ways by Boyle, Hales, Lemery, Boerhaave amongst others. See Dyck [76].

[25]On the history of specific and latent heats, see Scott [78], Dyck [79], Fox [80], McKie and Heathcote [81].

latter. Affinity theories were extended on the assumption that they were a solid foundation even as the phenomena of heat eroded that very assumption. One line of defence was to turn affinity from a qualitative theory into a quantitative one, enabling the true mathematisation of the discipline. Attempts were made to quantify affinity by Richard Kirwan, who developed a theory that broadly equated the strength of affinity between two substances with their combining proportion, and much later, in the 19th century by Thomas Young who seems to have used a rather more arbitrary system [26, 84–87]. Indeed it seems that Lavoisier himself was, before his unfortunate end, contemplating a line of enquiry that might have promised a quantified affinity, and a truly predictive science [88, 89].

Affinity theories thus increased in both explanatory power and heuristic scope, reflecting the disciplinary development of chemistry. In the later years of the century, affinity held the privileged position of being at once a black boxed instrument of the greatest utility to chemists and underpinning their discipline, and at the same time an object of investigation in itself.

4.5 Revolution

The spectre of phlogiston haunts any discussion of 18th century chemical theory and is as difficult to exorcise from historical writing about the period as Banquo's ghost from Macbeth's banquet. As the example of Cullen's speculations on the nature of phlogiston suggests, however, for many practitioners for much of the century it was no more vital a part of the content of the science than was any other substance. For Cullen, and presumably for many of his students and contemporaries, the idea of a 'phlogiston' was a fluid concept, to be toyed with and speculated upon, but by no means to be regarded dogmatically or as an essential part of their science. From the earliest days of Geoffroy's original table, phlogiston had featured, in some cases just as one substance amongst others in the column of another (as, for instance, in Geoffroy's table it was included as 'soufre principe', a substance that he claimed combined with vitriolic acid to form sulphur[26]). In other

---

[26]It is possible, of course, that the real point of Geoffroy's paper was not affinity at all. Some years earlier he had reported a series of experiments which he interpreted as showing that a sulphur principle combined with vitriolic acid to form common sulphur and with metallic calces to form the more familiar metals [90, 91]. Bearing in mind that the former combination is specifically included in his affinity table, perhaps the 1718 *Mémoire* was in fact a piece of subtle propaganda intended to reinforce his views? In a later paper that defended certain aspects of his table against criticism, he identified the sulphur principle with Stahl's phlogiston [92]. Here he expressly focused on its combination with vitriolic acid as an instance of an affinity that was stronger than that of the acid for fixed alkali. His paper reinforced his interpretation of his 1704 experiments even as he was purportedly clarifying his rapports. These repeated references to his sulphur principle were embedded within papers that drew almost entirely on familiar, even unquestioned knowledge. Both papers thus implied that his new ideas were not only accepted, they were beyond question. This idea is, it must be admitted, purely speculative, although it usefully illustrates the point perhaps, that Geoffroy's own intentions for his table were irrelevant to its subsequent history.

tables, such as those prepared by Bergman, phlogiston appears at the head of its own column, showing its affinities for other substances such as acids and metallic calces. In the second edition of his *Dissertation on Elective Attractions,* next to the column for phlogiston, Bergman also included a column for "matter of heat". He devised this column on the basis of observations of the flow of heat from the mercury in a thermometer when placed in an evacuated air pump. He apparently understood these phenomena as proceeding from the successive attraction of heat by these different substances in a fashion a little like capillary action [93]. When water was placed on the bulb of an open thermometer and the air pump evacuated, the heat capacity of the rarefied air increased, drawing heat from the water. The water in turn attracted heat from the glass of the thermometer, and this finally drew heat from the mercury which contracted as a result and fell in the glass. The attraction of air for heat was thus presumed to be stronger than for water, the attraction of heat for glass was weaker and for mercury weaker still. Bergman believed that specific heats were a product of the particular attraction of a body for heat, and what he called the external and internal surfaces. These corresponded generally to the internal and external surface areas of bodies, the porous structure of which he compared to a sponge. As bodies changed state, their internal surface area changed accordingly (the greater the bulk, the larger the internal surfaces). Thus physical state and attraction combined produced specific heat [94].

Bergman's affinity tables took on their own ghostly afterlife well into the 19th century, long after the chemical revolution. The post-Lavoisierian British chemist George Pearson (a student of Joseph Black, himself a student of William Cullen) amended and extended Bergman's table to take account of the new chemistry [95].[27] He reissued his translation of the *Nomenclature* alongside his new tables which were "from *Bergman,* with alterations and additions" [95]. Pearson adopted Bergman's matter of heat column, renaming it 'calorific or gasogen' and adding at the bottom the "bases of all the gases", presumably in concordance with Lavoisier's theory. Needless to say, Bergman's column for phlogiston has no parallel in Pearson's table, although it is interesting to note that he omits to include any column for hydrogen. Pearson's tables were picked up almost immediately and used in chemical reference works such as Nisbet's *General Dictionary of Chemistry* [96] and Parkinson's *Chemical Pocketbook* [97].

In the latter work, Parkinson claimed that caloric combined chemically with matter according to its affinity for each particular substance.[28] The "combined caloric" was explicitly equated with Black's latent heat, while heat capacity was proportional to the affinities between caloric and matter. Pearson's dephlogisticated version of Bergman's table appeared alongside this theory with a column for 'calorific' which was still predominantly based on Bergman's idiosyncratic method

---

[27]Possibly the largest affinity table ever produced at 62 columns in both the wet and dry ways.

[28]Parkinson was scrupulous in presenting alternative views from the material view of heat as well. His personal preference can, however, be ascertained by the fact that the four pages devoted to the caloric theory was 'balanced' by an account of Rumford, Davy and Beddoes's arguments in favour of vibratory heat occupying barely a single page.

of ordering which accorded with the heat capacities that might be expected from Parkinson's theory.

What is clear, is that the doctrine of affinity survived the chemical revolution largely unscathed. A few columns of the affinity tables were renamed, a few were removed, and a few were added. In most cases, once the names had been changed, the columns required no more amendment than would have been commonplace in the earlier decades. Lavoisier's own *Traité Elementaire de Chimie* included tables "dans l'ordre de leur affinité avec cet acide" etc.[29] Similarly, the book often regarded as the playbook for the chemical revolution, Richard Kirwan's (hijacked) *Essay on Phlogiston and the Constitution of Acids* saw both 'sides' of the revolution arguing their case on the basis of affinity [99]. Kirwan's portion of the *Essay* drew on affinity theory to point out the inconsistencies of Lavoisier's claims, pointing out that they did not accord with what he had presented as the affinities of the oxigenous principle. Similarly, Lavoisier's supporters responded to Kirwan's criticisms by referring to affinity in support of their own arguments. Affinity was here set forth in terms evoking Cullen's 'general principle' or Macquer's 'fundamental truth'. In this context affinities were regarded as axiomatic; if a theory contradicted the observed affinities, then the theory must be at fault.

The to-ing and fro-ing between Kirwan and the French chemists in the *Essay on Phlogiston* emphasise that this point of view was shared by both sides of the debate. Some historians have argued that the key to the success of the oxygen theory was the new tactic of providing an entirely new nomenclature, use of which entailed an implicit subscription to the new theory. It might also be added that Lavoisier's strategy also relied on retaining a familiar foundation beneath his new edifice, and that his chemistry, however new, remained intelligible, and thereby more likely to be adopted, due in part to his continuing references to and uses of an unchanged affinity theory. The substances might have new names, but what they were doing was familiar, and couched in familiar terms.

Some years later, Thomas Thomson wrote [62]:

> Though chemical affinity constitutes confessedly the basis of the science, it had been almost completely overlooked by Lavoisier, who had done nothing more on the subject than drawn up some tables of affinity, founded on very imperfect data.

In fact, as Beretta has shown, Lavoisier's avoidance of affinity theory is a fallacy [88]. A memoir entitled "Vues Générales sur le Calorique" published in 1805 by Madame Lavoisier together with other previously unpublished work shows that Lavoisier did consider affinity as of vital importance to his chemistry [89]. In this memoir he unambiguously related affinity to universal gravitation, linking it, as Parkinson had (although in a different way) to his caloric theory of heat.[30]

---

[29]Lavoisier [98], see Tables of Contents and throughout.

[30]Briefly, the particles of caloric tended to separate particles of matter (molecules), while the force of attraction pulled them together. The states of matter and its behaviour were a consequence of the balance of these two forces. Thus, for Lavoisier the two "pillars" were presumably caloric and affinity: this would seem to cast some doubt on the true extent of Lavoisier's 'revolution' [89].

The details of his theory are beyond the scope of this chapter, but it is clear that affinity was considered, by Lavoisier as well as Kirwan, to form the basis of chemistry, with or without phlogiston.

4.6 Conclusion

When we come to think about theory choice in eighteenth century chemistry, we have to reluctantly acknowledge the existence of a large grey wrinkled pachyderm which, try as we might, we cannot hide behind even the largest screen. Thomas Kuhn settled the chemical revolution firmly at the heart of his philosophy of science as the archetype of scientific revolution; we might even say the very paradigm of paradigm shifts [100]. The effect of this, as modern historians of 18th century chemistry will be found arguing at many a conference dinner, is a notable tendency amongst popular historians and many philosophers of science to view the 18th chemistry as dominated by phlogiston and its 'overthrow', almost to the exclusion of any other theoretical content. In the hands of even worse informed popular science writers, this becomes an implication that chemistry before the chemical revolution was some kind of pseudo-science, the occupation of the credulous and ignorant, and that everything prior to revolution was thankfully binned from 1789. Hence the kind of bizarre statement encountered in a recent (very) popular science book that Lavoisier's *Traité* "was considered the first true chemistry book" [101]. Kuhn's characterisation of the chemical revolution as 'paradigm shift' has not always been of assistance in the production of good history.

As we have seen, the Kuhnian model is problematic when examining the century as a whole. Affinity theory, and the affinity tables that were omnipresent by mid-century became so not because of any revolution, but because they were pedagogically useful, and in the pedagogical context so often ignored by Kuhnian philosophers and historians, a useful and progressive discipline was forged. Perhaps we might characterise this in Kuhnian terms, by stating that affinity became part of normal science. At the same time, however, this was a discipline in constant flux, a discipline that was yearly discovering new substances, elements, compounds, metals and most of all, new 'airs'; pre-revolutionary chemists may not have appreciated that water was a compound, but they were nevertheless capable of generating new and important knowledge using their affinity theories to guide them.

Similarly, once we follow affinity theory up to the metaphorical brick wall of the chemical revolution, we can see that while phlogiston may not have been able to scale the wall, affinity, more fundamental to chemistry, and more flexible due in part to its sheer vagueness, simply wriggled under. To try another metaphor on for size, while the paradigm may have shifted and while the chemists on the far side of the revolution may have seen a rabbit rather than a duck, the page on which the rabbit/duck was inscribed had not been turned. Contrary to much of the historiography which prefers revolution to normality, the discipline was not all about phlogiston, any more than it was all about sulphur, or heat, or oxygen/

dephlogisticated air. It was however (and for much of the early 19th century continued to be), about the interactions between different kinds of matter and the products of these different kinds in combination. If the kinds changed, as they had been doing over the previous centuries, as many chemists fully expected them to continue doing, that was not problematic. The doctrine of affinity provided a heuristic that guided chemists in their analysis of new substances, enabling the new to be understood in similar terms to the familiar. New columns could be added, old ones divided, amended or removed, but the concept of affinity remained, and would outlast many of the new chemical kinds.

References

1. Thackray A (1970) Atoms and powers, harvard monographs in the history of science. Oxford University Press, London
2. Thackray A (1995) Quantified chemistry—the Newtonian dream. Osiris 2nd series, 10:92–108
3. Cohen IB (1964) Isaac Newton, Hans Sloane and the Académie Royale des Sciences. In: Melanges Alexandre Koyré - 1 L'Aventure de la Science, vol. 1, Histoire de la Pensee - Ecole Pratique des Hautes Etudes, Sorbonne. Hermann, Paris, pp 61–116
4. Crosland M (1959) The use of diagrams as chemical 'equations' in the lecture notes of William Cullen and Joseph Black. Ann Sci 15:75–90
5. Crosland M (1963) The development of chemistry in the eighteenth century. In: Transactions of the first international congress on the enlightenment, Besterman, Theodore. Studies on Voltaire and the Eighteenth Century, vol. 24. Institute de Musee Voltaire, Geneva, pp 369–441
6. Hankins TL (1985) Science and the enlightenment. Cambridge University Press, Cambridge
7. Geoffroy EF (1719) Table des Differents Rapports observes en Chimie entre differentes substances. Mem Acad R Sci 1718:202–212
8. Fontenelle BB (1719) Sur les Rapports de Differentes substances de Chimie. Mem Acad R Sci 1718:35–37
9. Duncan AM (1981) Styles of language and orders of chemical thought. Ambix 28:83–107
10. Duncan AM (1996) Laws and order in eighteenth-century chemistry, chapters 3 & 4. Clarendon Press, Oxford
11. Académie Royale des Sciences (1962) Proces Verbaux, 1667–1793. Num. BNF de l'éd. de, Paris, T37, f 231v
12. Fontenelle BB (1722) Sur les Rapports Des Differents Substances en Chimie. Mem Acad R Sci 1720:32–35
13. Senac JB (1723) Nouveau Cours de Chymie Suivant les Principes de Newton & de Sthall. Paris, p lxvii
14. Duncan AM (1996) Laws and order in eighteenth-century chemistry. Clarendon Press, Oxford, p 114
15. Newton I (1979) Opticks or a treatise of the reflections, refractions, inflections & colours of light. Dover Publications, Inc., New York, pp 380–381 (reprint of 1730 edition)
16. Taylor GNL (2006) Variations on a theme: patterns of congruence and divergence among 18th century chemical affinity theories. Ph.D. thesis, University College London
17. Guerlac H (1968) The Background to Dalton's Atomic Theory. In: Cardwell DS (ed.) John Dalton and the Progress of Science. Manchester University Press, Manchester, p 73
18. Boyle R (1744) Of the mechanical causes of chemical precipitation (1675). In: The works of the honorable Robert Boyle, 5 vols, vol 3. A. Millar, London, pp 635–642

19. Boyle R (1744) Of the mechanical causes of chemical precipitation (1675). In: The works of the honorable Robert Boyle, 5 vols. A. Millar, London, vol 3, p 640
20. Kragh H (1989) An introduction to historiography of science. Cambridge University Press, Cambridge, p 111
21. Warltire J (1769) Tables of the various combinations and specific attractions of the substances employed in chemistry. Being a compendium of that science: intended chiefly for the use of those gentlemen and ladies who attend the Author's lectures. London, p 25
22. Cullen W (n.d. 1760s?) Notes taken from Chemistry Lectures. MSS 10, Cullen, Royal College of Physicians Edinburgh Library, Edinburgh
23. Kim MG (2003) Affinity, that elusive dream: a genealogy of the chemical revolution. Transformations: studies in the history of science and technology. Cambridge University Press, Cambridge, Massachussetts, pp 188–201
24. Kim MG (2003) Affinity, that elusive dream: a genealogy of the chemical revolution. Transformations: studies in the history of science and technology. Cambridge University Press, Cambridge, Massachussetts, p 167
25. Geoffroy EF (1736) A treatise of the fossil, vegetable and animal substances that are made use of in physick (trans: Douglas G). London
26. Taylor G (2008) Marking out a disciplinary common ground: the role of chemical pedagogy in establishing the doctrine of affinity at the heart of British chemistry. Ann Sci 65:465–486
27. Gibbs FW (1960) Itinerant lecturers in natural philosophy. Ambix 8:111–117
28. Saunders W (n.d. 1766?) A syllabus of lectures on chemistry
29. Martin B (1743) A course of lectures in natural and experimental philosophy, geography and astronomy. Newbery and Micklewright, Reading
30. Shaw P (n.d. 1733?) Chemical Lectures, Publickly read at London, in the years 1731, and 1732; and Since at Scarborough, in 1733 for the improvement of Arts, Trades and Natural Philosophy. London
31. Warltire J (1769) Analysis of a course of lectures in experimental philosophy; with a brief account of the most necessary instruments used in the course, and the gradual improvements of science: intended chiefly for the use of the Author's audience, 6th Ed. London
32. Wilson J (1771) A course of chemistry divided into twenty-four lectures, formerly given by the late learned Doctor Henry Pemberton, Professor of Physic at Gresham College
33. Golinski J (1983) Peter Shaw: Chemistry and Communication in Augustan England. Ambix 30:19–29
34. Shaw P, Hauksbee F (1731) An Essay for Introducing a Portable Laboratory: By means whereof all the Chemical Operations are Commodiously Perform'd for the Purposes of Philosophy, Medicine, Metallurgy, and a Family. J. Osborn and T. Longman, London, p 41
35. Shaw P (n.d. 1733?) Chemical Lectures, Publickly read at London, in the years 1731, and 1732; and Since at Scarborough, in 1733 for the Improvement of Arts, Trades and Natural Philosophy. London, pp 173–174
36. Boerhaave H (1727) A new method of chemistry. In: Shaw P, Chambers E. (eds) J. Osborn and T. Longman, London
37. Powers JC (2012) Inventing chemistry: Herman Boerhaave and the reform of the chemical arts. The University of Chicago Press, Chicago
38. Boerhaave H (1741) A new method of chemistry. In: Shaw P (ed), London, p 58
39. Lewis W (1748) Proposals for Printing, by Subscription Commercicum Philosophico-Technicum, frontispiece
40. Lewis W (1763) Commercicum Philosophico-Technicum or The Philosophical Commerce of Arts: designed as an attempt to improve Arts, Trades and Manufactures, 2 vols. London, frontispiece
41. Lewis W (1748) Proposals for Printing, by Subscription Commercicum Philosophico-Technicum, p 18
42. Lewis W (1763) Commercicum Philosophico-Technicum or The Philosophical Commerce of Arts: designed as an attempt to improve Arts, Trades and Manufactures, 2 vols. London, p iv

43. Lewis, W (1753) The New Dispensatory: Containing I. The Theory and Practice of Pharmacy. II. A Distribution of MEDICINAL SIMPLES, according to their virtues and sensible Qualities; the Description, Use, and Dose of each Article. III. A full translation of the LONDON and EDINBURGH PHARMACOPOEIAS, with the Use, Dose &c. of the several Medicines. IV. Directions for EXTEMPORANEOUS PRESCRIPTION; with a select number of elegant FORMS. V. A collection of CHEAP REMEDIES for the Use of the POOR. The Whole Interspersed With Practical Cautions and O Observations. Intended as a CORRECTION, and IMPROVEMENT of Quincy, 1st edn. J. Nourse, London, p 10

44. OED (2004) Oxford University Press. http://dictionary.oed.com/cgi/entry/00330510. Accessed 10 Oct 2006 (Online)

45. Sivin N (1962) William Lewis (1708–1781) as a Chemist. Chymia 8:63–88

46. Lewis W (1765) The new dispensatory, 2nd edn. London, p 35

47. Clow A, Clow N (1952) The chemical revolution. The Batchworth Press, London, pp 165–177

48. Bensaude-Vincent B (1986) Mendeleev's periodic system of chemical elements. Br J Hist Sci 19:3–17

49. Cullen W (1766) Notes taken by Charles Blagden from Chemistry Lectures, MS 1922, Blagden Papers, Wellcome Library for the History and Understanding of Medicine, London, Lecture 9

50. Cullen W (1766) Notes taken by Charles Blagden from Chemistry Lectures, MS 1922, Blagden Papers, Wellcome Library for the History and Understanding of Medicine, London, Lecture 19

51. Cullen W (n.d. 1760s?) Notes taken from Chemistry Lectures. MSS 10, Cullen, Royal College of Physicians Edinburgh Library, Edinburgh., f 66

52. Duncan AM (1996) Laws and order in eighteenth-century chemistry. Clarendon Press, Oxford, p 115

53. Cullen W (1766) Notes taken by Charles Blagden from Chemistry Lectures, MS 1922, Blagden Papers, Wellcome Library for the History and Understanding of Medicine, London, Lecture 10

54. Macquer P-J (1749) Elemens de Chymie—Theorique. Chés Jean-Thomas Herissant, Paris, p 256

55. Macquer P-J (1749) Elemens de Chymie—Theorique. Chés Jean-Thomas Herissant, Paris, p 22

56. Kim MG (2003) Affinity, That elusive dream: a genealogy of the chemical revolution. Transformations: studies in the history of science and technology, chapter 4. Cambridge University Press, Cambridge, Massachussetts

57. Kim MG (2003) Affinity, that elusive dream: a genealogy of the chemical revolution. Transformations: studies in the history of science and technology. Cambridge University Press, Cambridge, Massachussetts, p 161

58. Klein U (2001) Tools and modes of representation in the laboratory sciences. U. Boston studies in the philosophy of science. Kluwer Academic Publishers, Dordrecht

59. Klein U (2001) Berzelian formulas as paper tools in early nineteenth century chemistry. Found Chem 3:7–32

60. Klein U (2001) The creative power of paper tools in early nineteenth-century chemistry. In: Klein U (ed) Tools and modes of representation in the laboratory sciences. Boston studies in the philosophy of science. Kluwer Academic Publishers, Dordrecht, pp 13–34

61. Anderson RGW, Jones J (2012) The correspondence of Joseph Black, 2 vols. Ashgate, Farnham

62. Thomson T (1830) The history of chemistry, 2 vols. Henry Colburn and Richard Bentley, London, p 157

63. Duncan AM (1996) Laws and order in eighteenth-century chemistry. Clarendon Press, Oxford, p 116

64. Klein U (1996) The chemical workshop tradition and the experimental practice: discontinuities within continuities. Sci Context 9(3):251–287

65. Klein U (1995) E F Geoffroy's table of different rapports observed between different chemical substances—a reinterpretation. Ambix 42:79–100
66. Galison P, Assmus A (1989) Artificial clouds, real particles. In: Gooding D, Pinch TJ, Schaffer S (eds) The uses of experiment: studies in the natural sciences. Cambridge University Press, Cambridge, pp 225–274
67. Read J (1995) From alchemy to chemistry. Dover Publications Inc, New York, p 79
68. Cobb C, Goldwhite H (2001) Creations of fire: chemistry's lively history from alchemy to the atomic age. Perseus Publishing, Cambridge, Massachusetts
69. Lewis W (1753) The New Dispensatory: Containing I. The Theory and Practice of Pharmacy. II. A Distribution of MEDICINAL SIMPLES, according to their virtues and sensible Qualities; the Description, Use, and Dose of each Article. III. A full translation of the LONDON and EDINBURGH PHARMACOPOEIAS, with the Use, Dose &c. of the several Medicines. IV. Directions for EXTEMPORANEOUS PRESCRIPTION; with a select number of elegant FORMS. V. A collection of CHEAP REMEDIES for the Use of the POOR. The Whole Interspersed With Practical Cautions and O Observations. Intended as a CORRECTION, and IMPROVEMENT of Quincy, 1st edn. J. Nourse, London, p 7
70. Lewis W (1753) The New Dispensatory: Containing I. The Theory and Practice of Pharmacy. II. A Distribution of MEDICINAL SIMPLES, according to their virtues and sensible Qualities; the Description, Use, and Dose of each Article. III. A full translation of the LONDON and EDINBURGH PHARMACOPOEIAS, with the Use, Dose &c. of the several Medicines. IV. Directions for EXTEMPORANEOUS PRESCRIPTION; with a select number of elegant FORMS. V. A collection of CHEAP REMEDIES for the Use of the POOR. The Whole Interspersed With Practical Cautions and O Observations. Intended as a CORRECTION, and IMPROVEMENT of Quincy, 1st edn. J. Nourse, London, p 12
71. Cullen W (1766) Notes taken by Charles Blagden from Chemistry Lectures, MS 1922, Blagden Papers, Wellcome Library for the History and Understanding of Medicine, London, Lectures 40–43
72. Anderson R (2005) Joseph Priestley: public intellectual. Chem Herit Mag 23(1):6–9, 36–38
73. Bergman T (1970) A Dissertation on elective attractions, 2nd English edn (trans: Beddoes T) Frank Cass & Co. Ltd., London (reprint of 1785 edition)
74. Nicholson W (1785) A dictionary of chemistry. G. G. and J. Robinson, London, p 158
75. Nicholson W (1792) The first principles of chemistry, 2nd edn. G. G. and J. Robinson, London, p 75
76. Dyck DR (1967) The Nature of heat and its relationship to chemistry in the eighteenth century, chapters II and IV. Ph.D. thesis, University of Wisconsin
77. Newton I (1979) Opticks or a treatise of the reflections, refractions, inflections and colours of light. Dover Publications, Inc., New York, pp 377–378 (reprint of 1730 edn)
78. Scott EL (1981) Richard Kirwan, J H de Magellan, and the early history of specific heat. Ann Sci 38:141–153
79. Dyck DR (1967) The nature of heat and its relationship to chemistry in the eighteenth century. Ph.D. thesis, University of Wisconsin
80. Fox R (1971) The caloric theory of gases from Lavoisier to Regnault. Clarendon Press, Oxford
81. McKie D, Heathcote NHV (1935) The discovery of specific and latent heats. Edward Arnold, London
82. Fox R (1971) The caloric theory of gases from Lavoisier to Regnault. Clarendon Press, Oxford, pp 19–20
83. McKie D, Heathcote NHV (1935) The discovery of specific and latent heats. Edward Arnold, London, p 137
84. Kirwan R (1781) Experiments and observations on the specific gravities and attractive powers of various saline substances. Phil Trans R Soc Lond 71:7–41
85. Kirwan R (1782) Continuation of the experiments and observations on the specific gravities and attractive powers of various saline substances. Phil Trans R Soc Lond 72:179–xxxv

86. Kirwan R (1783) Conclusion of the experiments and observations concerning the attractive powers of the mineral acids. Phil Trans R Soc Lond 73:15–84
87. Young T (1809) A numerical table of elective attractions; with remarks on the sequences of double decompositions. Phil Trans R Soc Lond 99:148–160
88. Beretta M (2001) Lavoisier and his last printed work: the Mémoires de Physique et de Chimie (1805). Ann Sci 58:327–356
89. Lavoisier AL (2004) Mémoires de Physique et de Chimie, 2 vols. Thoemmes Continuum, Bristol, vol I, pp 1–28 (reprint of 1805 edition)
90. Geoffroy EF (1704) Maniére de Recomposer le Souffre Commun par la Réünion de ses Principes, et d'en composer de Nouveau par le Mélange de Semblables Substances, avec quelques conjectures sur la composition des Métaux. Mem Acad R Sci 278–286
91. Geoffroy EF (1709) Expériences sur les Métaux, faites avec le Verre ardent du Palais Royal. Mem Acad R Sci 162–176
92. Geoffroy EF (1720) Enclairissements Sur la Table inferée dans les Mémoires de 1718 concernant les Rapports observés entre differentes Substances. Mem Acad R Sci 20–34
93. Bergman T (1970) A Dissertation on Elective Attractions, 2nd English edn. (trans: Beddoes T) Frank Cass & Co. Ltd., London, pp. 240–248 (reprint of 1785 edition)
94. Bergman T (1970) A Dissertation on Elective Attractions, 2nd English edn. (trans: Beddoes T) Frank Cass & Co. Ltd., London, pp 237–139 (reprint of 1785 edition)
95. Anon [Pearson G] (1799) A Translation of the Table of Chemical Nomenclature Proposed by De Guyton, Formerly De Morveau, Lavoisier, Bertholet, and De Fourcroy with Explanations, Additions, and Alterations to which are subjoined Tables of Single Elective Attraction, Tablers of Chemical Symbols, Tables of the Precise Forces of Chemical Attractions' and Schemes and Explanations of Cases of Single and Double Elective Attractions, 2nd edn. J Johnson, London, Table III
96. Nisbet W (1805) A general dictionary of chemistry, containing the leading principles of the science in regard to facts, experiments and nomenclature, for the use of students
97. Parkinson J (1800) A chemical pocketbook. London, p 6
98. Lavoisier A (1789) Traité Élémentaire de Chimie. Paris
99. Kirwan R (1968) An Essay on Phlogiston and the Constitution of Acids. Frank Cass & Co Limited, London (reprint of 1789 edition)
100. Kuhn T (1996) The structure of scientific revolutions, 3rd edn. University of Chicago Press, Chicago
101. Flynn S (2012) The science magpie. Icon Books, London, p 77

Chapter 5
One of These Things is Just Like the Others: Substitution as a Motivator in Eighteenth Century Chemistry

Matthew Paskins

5.1 Introduction

In their recent book *Materials in Eighteenth Century Science*, Ursula Klein and Wolfgang Lefèvre argue that eighteenth century chemistry was primarily a science based upon the knowledge of materials. Materials and substances were [1]:

> multidimensional objects of inquiry that could be investigated in practical and theoretical contexts and that amalgamated perceptible and imperceptible, useful and philosophical, technological and scientific, social and natural features.

This had strong theoretical consequences. Because the imperceptible features of chemical substances were not the only relevant consideration, plural ontologies and theoretical commitments could exist within cultures of chemical study, and perceptible features—the taste, smell and other forms of knowledge which we associate with the 'sensuous' chemist—retained primary significance.[1] So, for many materials, did descriptions on the basis of provenance, though in the reforms of nomenclature and taxonomy many substances came to classified on the basis of their chemical composition, not their other properties.

This work is an enormously suggestive account of meta-theoretical choices faced by the chemists of this period. But it has two features which—in terms of the broader history of scientific materiality—are striking and potentially problematic.

First, chemistry is heavily privileged as the science of materials [1]:

---

[1]For which, see Roberts [2].

M. Paskins (✉)
"Commodity Histories" Project, Open University, Milton Keynes MK7 6AA, UK
e-mail: matthew.paskins.09@ucl.ac.uk

© Springer International Publishing Switzerland 2016
E. Tobin and C. Ambrosio (eds.), *Theory Choice in the History of Chemical Practices*, SpringerBriefs in History of Chemistry,
DOI 10.1007/978-3-319-29893-1_5

'learned inquiries into materials took place in mineralogy, botany, pharmacy, architecture, and a few other areas', but 'chemistry was the only scientific culture in the eighteenth century where materials were studied persistently, comprehensively, and from multiple perspectives.'

This is a disciplinary land-grab. It is intended to provoke questions about the relations between science technology, natural history and natural philosophy. It can be read in a strong or weak sense, depending on what we take the disciplinary boundaries of chemistry to be. In its strong sense, only the practices of taxonomy plus laboratory practice which provide the persistent, comprehensive and plural enquiry which Klein and Lefèvre seek to describe. They are certainly open to the possibility that chemistry, in some moments 'collaborated with' natural history [3]. In a weaker sense, 'chemistry' was open to alternative practices, such as the descriptions associated with botanical natural history; it was this openness which gave chemistry its special concern with materials during this period. Either way, the science is privileged—over unanalyzed alternative approaches—in its power to define and organize the properties of materials.

This leads to the second problem. The work is oriented towards schemes of classification, chemists' orderings of materials, rather than descriptions of the treatment of materials themselves. Because Klein and Lefèvre are writing about significant transitions in treatments of materials and the systems according to which they can be known, this approach is well-motivated. Despite all the categorical blurrings which it entails, and the allusions to society and commodity as embedded qualities within materials, it continues to privilege the taxonomic and analytic procedures of science as sources of knowledge.[2]

Other material histories have taught us different analytic and descriptive tricks. In particular, students of commodity history have employed a version of 'bundle theory', which finds its pedigree in the writings of David Hume. According to this view, all materials are bundles of different properties, which may be incoherently related to each other: my apple is red, costs 79 pence, has the texture of mulched paper, was grown in Kent without pesticides, and provides the home for a maggot of surprising vitality. No science has any particular privilege in defining the properties we might ascribe to an object. We don't need to ascribe any great ontological weight to bundle-theory to use it in commodity history. What it always relevantly entails is that any given material may have one or more properties which in which we are interested, and others which don't concern us. These other properties might turn out to be relevant at a later date (most obviously in the case that the material turns out to be harmful).

Bundle theory becomes most interesting when we think about a particular way of dealing with materials: substituting new materials for others of high value. This was a major project, across a wide range of sciences during the eighteenth century: there were powerful economic and political rationales for substitution, in the hope of

---

[2]Thus Klein and Lefevre offer a remarkably serene revision to Foucault's deeply problematic schematisation in Les Mots et Les Choses [4].

replacing expensive or scarce imports, for example, or developing your colonial economy or proving that Uppsala was paradise and could well-sustain a pineapple industry.[3] Bundle theory becomes an interesting approach to talking about substitutive efforts because substitution requires a fine determination of which properties are and are not relevant for us in a material. According to Maxine Berg, imitation of oriental luxuries in European materials—which didn't behave exactly as expected—led directly to product innovation. She calls this process 'imitative invention', and it comes down to the surprising extra properties of materials, which we had initially considered to be irrelevant [7, 8].

In the rest of this paper I want to pick up the idea of substitution as ways of understanding how properties are bundled into materials, and use it to build on Klein and Lefèvre view about the importance of materials in eighteenth century chemistry. Section two describes the possibilities and challenges of substitution for eighteenth century chemical practice, emphasizing the combination of analytical and natural descriptive approaches to tackling problems where substitution was possible. This fleshes out the idea that materials were multi-dimensional objects of enquiry by showing how different strategies—of purification, same-making, and imitative invention—were used by chemists concerned with materials. Section three picks up the question of the relation between analytic and descriptive techniques; this tries to clarify the balance of powers between the two, which Klein and Lefèvre have left vague. Drawing on this account of substitution, section four offers a detailed revision of a standard case-study which puts analytic chemistry on top and descriptive practices at the bottom. This study in question is James Delbourgo's account of the work of the London spy, natural historian and dyer Edward Bancroft [9]. Delbourgo misreads Bancroft's institutional place and the orientation of his chemistry because he takes too seriously the rhetoric of his prefaces which sets 'chemical culture' against other forms of knowledge and practice. The final section concludes.

5.2   Substitution

In 1784, a paper appeared in the *Philosophical Transactions of the Royal Society,* written by none other than James Watt [10]. It concerned the use of indicators other than litmus for detecting the acidity or alkalinity of chemical substances. Litmus, prepared from lichens, had been known since the fourteenth century, but was generally accepted as the readiest indicator for chemistry, having replaced the 'syrop of violets', which, Watt wrote, was now regarded as insufficiently accurate. And for most purposes litmus was very accurate. But in some cases it was completely misleading [10]:

---

[3]For pointers, see Spary [5], Brockway [6].

a mixture of phlogisticated nitrous acid with an alkali will appear to be acid, by the test of litmus, when other tests, such as the infusion of the petals of the scarlet rose, of the blue iris, of violets, and of other flowers, will shew them same liquor to be alkaline, by turning green so very evidently as to leave no doubt.[4]

Watt tested several flowers and plants, and found most of those he tried rapidly lost their sensitivity, becoming useless in winter. So he looked among the winter vegetables, and decided that red cabbage was best. When it was fresh it had 'more sensibility both to acids and alkalies than litmus, and to afford a more decisive test, from its being naturally blue, turning green with alkalies, and red with acids; to which is joined the advantage of its not being affected by phlogisticated nitrous acid any farther than it acts as a real acid.' The rest of the article concerned techniques of extracting the colouring matter from the cabbage, and preserving its leaves from putrefaction [10].

As a moment in the history of indicators, this paper has some minor significance. Red cabbage had previously been used as an indicator by Robert Boyle, though Watt's rediscovery was independent of this use; it was later employed by Michael Faraday, though it is not clear how far Watt's own recipe was being followed [11].

The openness of the acidity test resembles to some extent the situation which existed in thermometry during the same period, as Hasok Chang has described it: chemists wanted to employ the properties of materials (say, the boiling point of mercury) to use as measures of other processes; but there was no independent way of obtaining those properties without drawing on your knowledge of other materials [12]. As Chang argues, this situation provides opportunities for iterative improvements in measurement technique rather than an infinite regress. Helpful kinds of comparison can be found. Less accurate procedures can be refined by comparison with others, and with known observable phenomena. This is exactly what was happening in the indicator case: Watt had the theoretical presumption that there was something suspicious about the very strong reaction between the litmus and the alkalized phlogisitcated acid, which made the response of the litmus seem like an anomaly, and opened the way to trying other indicators. That he considered the litmus in this way is underdetermined, but only very slightly: he could have chosen to trust the litmus, and regard the response of the others as the fact which needed explanation. But to do so would have involved rejecting the idea that acids were neutralized by alkalies, and that wasn't well-motivated.

What primarily interests me here, though, is the materials which Watt was employing for his test, and what they have to tell us about theoretical choices in eighteenth century chemistry. The cabbage and the rose petals were all locally produced and easily obtainable; they made appropriate comparators—and potential substitutes for—the litmus. All had visible properties which could be compared, without an over-arching analytic procedure for deciding how these responses worked. In fact, it was through these visible properties that analysis could be performed.

---

[4]Phlogisticated nitrous acid is now known as nitrous acid.

I'll begin by saying something about the promises and problems of eighteenth century substitution of materials, in a chemical context. It was common for chemists to speak explicitly about the viability of substituting local materials for ones which were more difficult to obtain. Watt was not explicitly using a rhetoric of material substitution, but it is clear that he was happy to try the physical-chemical properties of local flora by comparison with the standard test. The roots of this practice go back to antiquity and the Galenic *quid pro quo* tradition within *Materia Medica*.[5] There it was accepted that certain substitutions could be allowed for rarities. In eighteenth century books of practical chemistry, the *quid pro quo* idea survives but is given additional support by the idea that the analytical procedures of chemistry itself can inform on what materials are really the same as each other; and hence on a rational reduction in the number of different things which the *medica* contains.

For example, the third chapter of Robert Dossie's book *The Laboratory Laid Open* offered an [14]:

> examination of the sameness of several substances, which make a part of the *materia medica* under different denominations, without any essential diversity: being necessary for the determining, how far many substitutions are allowable.

Dossie cautioned against the idea that the provenance of a material was the cause of its chemically or medically significant properties, on the grounds that chemical purification of their salts showed that 'the supposed difference...betwixt salts of several kinds of vegetables, as of wormwood, broom, or tartar, does not consist, really, in the salts themselves', but rather through adulteration 'from the admixture of some other substance, with the salts': part of the oil of the vegetable. But in the analysis, such oil turned out to be extremely hard to separate from the original sample; and sometimes the admixture was exactly what you wanted.

Moreover, Dossie's optimistic rhetoric about finding the pure form of salts is belied by the relative rarity of such pure chemical substances across the whole range of eighteenth century chemical practice. This point is made in general terms by Klein and Lefevre. During the eighteenth century, they argue, pure chemical substances were those—primarily metals, alkalis, acids, earths—which were traceable in chemical transformations [15];

> in the making of salts and alloys, these traceable substances behaved like building blocks that were preserved in chemical transformations and could be recovered by decomposing the salts and alloys.

However [16],

> the bulk of raw materials and processed substances which seventeenth and eighteenth century chemists studied in their laboratories did not display this kind of recurrent pattern of chemical behavior. Their chemical transformations were for the most part much more difficult to trace.

---

[5]For the quid pro quo tradition, see Touwaide [13].

The problems of purity meant that other means of classification, analysis, and evaluation of properties were usually required. Analytical tests of materials were useful to learn what they were made from, and to detect adulterants, but could not discover the mix of properties which you hoped to use.

Between the 1750s and the 1790s, a large number of recipes appeared in English giving details of how to manufacture of potash, which was used in making lye for a range of bleaching, dyeing, and other chemical purposes. Alongside the recipes, some comprehensive efforts were made to compare the potash produced in different locales and establish which had the best properties. In 1756 the Scottish chemist Francis Home published his *Essays on Bleaching* which included a general account of different forms of available potash. He was surprised to report that the 'best' forms of potash—those most highly valued by bleachers—contained lime, which was illegal for use in Bleaching purposes in Britain and Ireland. Examining the different ashes comparatively, he concluded that [17]:

> There would appear, by my experiments, a greater difference than this betwixt the Swedish ashes, if that is the true process, [i.e. the recipe which Home had used] and those I have examined. I had discovered the greatest part of the Muscovy ashes to be lime. I suspected it might enter into the composition of the Marcroft and Cashub; and have accordingly discovered it there. *Without the same grounds, none would ever have searched for it.* Whence then comes this lime? It must either enter into its composition, or arise from the materials managed according as the process directs.

This was a ticklish point, because lime was prohibited from use in bleaching because it weakened cloth. What Home had demonstrated was that in ashes widely used for bleaching, lime *was* present—and it was the combination of lime with alkaline salts which allowed it to bleach without 'weakening and corroding' [18]. Some historians have claimed that Home's work led on to other researches which resulted in the repeal of anti-lime statutes [19]. Chronologically this doesn't make sense—the statutes remained in force until 1823, though various innovations had become possible in the meantime, including those which used lime as an ingredient in bleaching powder. Alongside the suggestion that lime might be a useful ingredient rather than a contaminant, Home discovered something else: a recipe for producing a local substitute for the Russian ashes, composed of one quarter of dissolved potash and three quarters of slaked lime. Half a century later, authors were unable to tell how far this particular recipe had spread [18].[6]

For the London Doctor John Mitchell, meanwhile, it was precisely the mixed-up nature of Swedish potash which made it peculiarly valuable. Mitchell noted that analysis had shown potash contained a 'metallic substance, which could be used in place of Prussian Blue dye; further 'the Combination of these principles makes many Properties in Pot-ash, more than what result from them in a State of Separation' [20]. Best of all were the crystallized salts which resembled nitre, and caused a degree of explosiveness when the potash was boiled—highly suitable for

---

[6]For the statutes see Statutes at Large …: (29 v. in 32) Statutes or the United Kingdom, 1801–1806; [1807–1832], p. 821.

the production of a soap, recommended by the College of Physicians and 'impregnated with their heating sulphureous quality'. Mitchell came up with a speculative chemical account of how the potash came by these qualities, but he was not interested in analysing it into its constituent parts; he alluded vaguely to the 'more volatile Salts of the Pine' which the Swedish process prevented from escaping.

Part of the motivation for these researches was the attempt to develop a potash industry in the north American colonies, which could compete with the high-quality products available from Sweden and Russia.[7] In response to concerns from the government of Massachussets that its potash was being adulterated, the London-based Society of Arts became involved in the development of tests for adulteration. One of these tests was a more precise form of titrimetry, developed by William Lewis. Using his titrimetric tests Lewis detected large amounts of lime and sea-salt in the potash samples and described them as adulterants. Dossie, who was also concerned with potash, but had a recipe which he had himself devised which he wanted to promote, reached an exactly opposite conclusion. What appeared to be adulterants were simply indications that the excellent process by which the potash had been prepared had been performed imperfectly. The Americans were honest but manufacture was hard. When it was done rightly, by contrast [23]:

> the American alkali, as prepared by the evaporation of the ley in vessels, according to the process published by the London Society of Arts [...] though at present called POT-ASH is not in reality of the same kind with the pot-ash, properly so called, made in Europe; but is, when the process has been rightly conducted, a pure fixt alkaline salt, free from any heterogenous matter: which the European pot-ash, as it is made by evaporation of the ley in the naked fire, can never be. [...] the American sort is now a different article of commerce from the European and must be judged of, as to its goodness and value, by a different standard of examination.

Dossie was playing fast and loose with the definition of 'European pot-ash'; he was describing that English and Irish process, not the Russian or Swedish. But his effort is interesting all the same: rather than accept a pre-existing standard of purity, he tried to claim that the newly substituted commodity was a new kind of product, which needed to be judged in new ways, rather than by the standards of purity which Lewis' test imposed.

In some ways these inventive efforts display similar iterative logic to Watt's use of the red cabbage as an indicator. To be sure, potash was not directly involved in processes of measurement or detection of material properties. But there was a back-and-forth between description of the material, specification of its properties, and inventive attempts to find ways of producing those properties in new forms, or in new commodities. Analysis of material composition certainly seems to have been helpful in some of these efforts, but so was the sense that some materials worked best because of the properties which they bundled together which interacted

---

[7]On these trials, see Stewart [21], Page [22].

together as more than the sum of their parts. As with iterative measurement, such imitative invention involved going from what could be known about materials on the basis of existing practice and finding ways to work from there.

5.3 Natural History and Analysis

On one level, there is nothing controversial or problematic about the kinds of procedure which I am describing here. The framework which I've been drawing on emphasizes taxonomic shifts in chemistry, and the presence of analytic alongside descriptive modes. What the previous section added was the inventive role of some kinds of chemical practice.

Historians of chemistry have long acknowledged that eighteenth century chemistry required a combination of two approaches, which correspond to 'inductive' as opposed to 'deductive' approaches, and correspond to Simon's 'data driven' and 'theory—driven' science or to John Pickstone's distinction between 'natural historical' and 'analytical' ways of knowing [24, 25].[8] According to Barbara Keyser's impressive account of Berthollet's contributions to the science of dyeing, the former involved 'careful selection and ordering of the facts of nature' which 'could reveal their significant relations and the general principles uniting them' [26]. Data-driven science, on the model of natural history was useful to specify what was in the 'problem domain'—the theory-driven approach then became useful when 'the problem domain [was] well understood.' Keyser describes this combination of two approaches, overall, as a 'double systematization' [26].

This double picture accounts for many of the issues which I have just described. 'Natural historical' descriptive approaches could be used to indicate the range of desirable properties which a chemical substance might have; this then allowed for the specification of precise analytical procedures to detect them and to allow decisions to be made about questions of chemical composition. For some chemical workers during the eighteenth century, it is quite clear, natural history which ordered materials according to their places of origin was a spur to develop new productive practices which could overturn existing hierarchies of value.

Thus the London steel-maker Henry Horne complained that a taxonomic list of different kinds of iron in *Chambers' Dictionary* had [27]:

> very arbitrarily assigned names and characters to different sorts of iron, according to the different countries where they are produced; this he has done in such a manner, (though without any real judgment), as to give the world a very high opinion of the iron of one country, to the great disparagement of that produced in another.

---

[8]According to Pickstone, natural history records 'what we've got'; analysis is anatomical. They overlap with and do not replace each other.

The historian Chris Evans, glossing this passage, argues that [27]:

> Horne argued instead for a form of classification in which the affinity of method rather than geographical provenance was paramount. Functional characteristics that were readily measurable were to take precedence over less tangible qualities like the "genius of the place".

Such readily measurable functional characteristics would seem to be open to attack through analytic procedures. But this was not always the case. Often—as in the case of potash—you wanted to make something in imitation of the best, where the material composition was not known in advance but a technique of production could be imitated. This imitative invention *could* involve analytic procedures for discovering the composition of materials, but it did not have to: all that mattered was that a material could pass—through whatever tests were available, as equivalent to the existing best. We saw this in the previous section with respect to potash; Horne's own judgments of the novel forms of steel which he produced always involved comparison to existing good varieties, such as Spanish steel.[9]

Building on this point, we need to be wary of an idealized, overly systemic picture of 'descriptive science' or natural historical ordering. Examples of such ideal taxonomic forms include the periodic table and many interpretations of the systemic work of Carl Linnaeus and his followers: divorced from local context, and providing their own way of shaping the data which they contained. Klein and Lefèvre's focus on chemical tables falls into this tradition of systemic description.

But eighteenth century natural history—broadly construed—was more quixotic, and less systematic, than is suggested by the idea of blithe general surveys of material properties.[10] This cashed out in theoretically-relevant ways which are particularly evident in attempted cases of substitution. They stemmed from the fact that, as Nicholas Jardine and Emma Spary put it, in natural history [29]:

> the boundaries between the natural and the conventional, artificial, and social have been continually contested and relocated.

The first was that natural history had its own techniques of trying to decide whether two things (plants, products, commodities) were equivalent in value, or were the same as each other. I will describe these procedures with respect to botany, which offered perhaps the most sophisticated approach to questions of identity over long distances.[11] But the outlines of the procedure can be given as follows.

---

[9]Horne also rejected existing chemical tests and experimental reports which told him that the materials he wanted to employ did not contain any iron.

[10]Historians of botany have, in general, significantly overstated the systemic nature of eighteenth century natural history—for the most forceful version of this story, see Foucault [28] and cp. Staffan Muller-Wille. Pickstone, by contrast, is keen to emphasise the descriptive and temporal aspects of natural history writing, a fact glossed over by Klein and Lefevre [4].

[11]For a worked-out example of the contingencies of this process—which does not really go beyond noting that it involves contingencies, see Emma Spary "Of Nutmegs and Botanists" [30]. My schematic is based on my detailed reading of the controversy between Philip Miller and John Ellis concerning the identities of various types of toxicodendron plants which appeared in the

In practice, to say that two useful plants were the same as each other was to claim (a) that they were botanically identical, and that this identity could be demonstrated on the basis of existing scholarly literature and on available samples of them (b) that even if their species were the same, they did not radically change according to the method by which they were cultivated and the location in which they were grown and (c) that there existed the potential for them to be exploited effectively in a new territory, as they were in an old.

'Natural history'—with respect to useful plants—was a portmanteau genre. It drew in evidence from travel accounts, descriptions of labour, botanical books, where known, the itineraries of seeds (to prove that a plant really had come from the location which was claimed), reports of direct employments of the plant's product in a process such as dyeing so that its material effects could be compared with those from other sources, and comparisons with substitute materials. It also involved reference to antiquarian studies, often going back to Pliny, descriptions of animal behavior, reports of committees and experiments, and so on. When they engaged in controversies, botanists with interests in properties of plants retranslated each other's sources, critiqued each other's drawings and insulted the regimes of management in each other's botanical gardens. They undermined or defended the claim to be able to identify a plant as the same through along multiple paths. And this was further complicated when the hope was to introduce the cultivation of a useful plant to a new location. For then the question of what was needed—in social and material terms—for the plant to grow and be used as well became a treacherous combination of botany, economics and speculation. This was a distinct way of knowing materials, and while chemistry could contribute to this brew, it was not the only ingredient, much less the master. In many of its manifestations, eighteenth century natural history was disorderly and historical: it did not provide nice sets of facts on which inductions could be performed.

5.4 Lac's Labours: Edward Bancroft's *Experimental Researches Concerning the Philosophy of Permanent Colours*

This section aims to flesh out the suggestions of the previous section about the unruly nature of natural history, and its role as a spur for chemical practice. I want to do this by giving a detailed reading of the work of a significant chemist who drew on natural historical principles. Besides his natural philosophical interests, Edward Bancroft was a spy and colonial go-between who wrote the *Natural History*

(Footnote 11 continued)

Philosophical Transactions between 1755 and 1758. Various commodity histories give interesting pointers on these themes; of these the best—if determinedly the oddest—is Robert Chenciner, Madder Red: A History of Luxury and Trade [31].

of Guiana and held a patent for importing Quercitron or black oak bark to Britain. In 1794, he published *Experimental Researches concerning the Philosophy of Permanent Colours*; a second edition appeared in 1813.

James Delbourgo, who has given an extensive account of Bancroft's achievements and subterfuges, concludes that the *Experimental Researches* was a celebration of European cultures of colour, identified with knowledge of the use of mordants [32]:

> "Savage tribes" might be masterful at changing the colour of skin, hair, feathers, and quills, but they were ignorant of the improvements philosophers had made through the use of mordants for greater permanence. […] Moving well beyond a Plinian account of the geographical provenance of materials, *Permanent Colours* constructed a world-historical hierarchy that ranked chemical explanation as the apex of human achievement in colour, above ancient arts and indigenous empiricism.

Finally, Delbourgo quotes a mid-nineteenth century treatise which praised Bancroft's work relatively highly, while noting that it was 'of little or no use in the dye-house, being too exclusively theoretical.'[12]

When we turn from these observations to the actual text of the *Experimental Researches,* however, the limitations of Bancroft's chemical hierarchy become clear. The work is primarily concerned with natural history and the provenance of dye-stuffs, not only analytic chemistry. I'll give a detailed example from chapter five of section two, on lacs. The section on 'lacs' compiles information and experiments from antiquarian and more recent sources on the behavior of lac insects and the appropriate ways for cultivating them. He quoted the great economic botanist William Roxburgh extensively to indicate how local Indian materials could be substituted for those used in equivalent processes elsewhere, in the preparation of the lake. He described its use by the 'natives of Assam'. There was much more parity in the details of the account between European and non-European practice than Delbourgo recognizes.

Bancroft's major goal was to establish 'the practicability of substituting the colour of the lac insect for that of cochineal' [34]. It is true that Bancroft employed chemical techniques to establish the properties of dye-stuffs: but this was theory as hunch-following and purification, not as material mastery. In 1797, he was given a [35]:

> parcel of colouring matter, which had very much the appearance of powdered cochineal, of which he gave me a few ounces, calling it East-Indian Cochineal, with a request that I would try its effects in dyeing scarlet.

Bancroft adapted a technique for 'receiving a scarlet from cochineal…by impregnating it with a muriatic solution of tin, and a certain portion of yellow colour from the quercitron bark'.[13] This procedure, which had been successful for

---

[12]Quoting E Parnell, A practical treatise on dyeing and calico-printing [33].

[13]Given Bancroft's patent on the quercitron bark, this may not have been an impartial part of the test.

cochineal from other places, didn't work; instead he added a little vegetable alkali, which produced a 'fine crimson colour' as 'the alkali' had 'separated the colouring matter from a portion of alumine which had been employed to precipitate it (in India), and to which it was too intimately united to be dissolved by water *only*. So he neutralized the resulting liquor and produced a good dye stuff, feeling [36]:

> full persuaded that the colouring matter which produced this effect was in reality nothing but the colouring matter of lac, extracted either when fresh, or by some particular means when dried, and afterwards precipitated either wholly or in part by alum.

There was controversy—with Berthollet—about whether lac was necessarily less bright than cochineal, or whether through combinations with 'preparations of tin' and other dyer's solvents [37].

The point is that the 'colouring matter' was the desirable product of all Bancroft's chemical tests, and was irreducibly associated with the productions of the beetles—the lac could imitate existing cochineal but chemical processes could not tell him anything about how the matter came to have the properties it did. Far from the hierarchical understanding posited by Delbourgo, the process remained an intolerable wrestle with powders and beetles. Any new substance to be tried out was an unknown from which chemical art might coax useful matter. Theories of the actions of dye-stuffs could then be employed in an ad hoc manner to try to persuade these principles to behave. And it needed to be through techniques which could be applied in the country of origin, a fact which bore on techniques of evaporation [38]:

> it being impossible in this climate, and at that season of the year, (which had been uncommonly wet,) to produce an extract by evaporation, with *only* the heat of the sun, and the aid of a dry atmosphere, as might be done in the East Indies, and, consequently, impossible by any experiment *here* to ascertain how far the preparation in question might be advantageously made in the country where *alone* I had proposed that It should be made, I resolved to the leave that question to be decided by *future* experiments in the East Indies.

Bancroft was subsequently invited to examine the imported commodity by the Court of Directors of the East India Company: these samples had been prepared by a Company surgeon at Keerpoy, prepared from fresh—rather than dried—lac, and kept in solution, and which had been favourably evaluated by the inspector of drugs for the board of trade in Bengal; Fleming had tried it himself [39]:

> and was agreeably surprised to find it, even under my inexpert management, to produce so good a colour. I have no doubt that in the hands of a skillful dyer it will bring out as bright and beautiful a scarlet as the cochineal itself.

The existing inferior stick lac was already used for dyeing red morocco leather, and had a healthy export market to Portugal and Barbary. However Bancroft was also active in trying to define a pure form of the lac, claiming that

> as it was prepared by different persons, with some variations in the quality, and as in all of it, the colouring matter was encumbered and deteriorated by other matters, partly extracted therewith, and partly added, to cause a precipitation of the colouring matter, so much difficulty and uncertainty attended the employment of the lac lake, that after having been

sold to profit for some time, it ceased to find purchasers, even at a fourth part of the price which it would have brought, if the colouring matter had not been so deteriorated; and the scarlet dyers in the year 1810 were generally determined to abstain from the use of it [40].

Bancroft had lost his records of his own experiments, but referred instead to Hatchett's "Analytical Experiments and Observations on Lac" which had been printed in the *Philosophical Transactions* of 1804. Here is what Hatchett found in the Lac: Colouring and other *animal* matters, composing or proceeding from the insects and their eggs [...] a resin very much resembling that produced by the hymenoea courbaril (commonly called gum anime) and that denominated copal, together with a small portion of a species of wax, possessing most of the properties of myrtle wax, obtained from the berries of the myrica cerfiera [41].

For these purposes, the resin was an adulterant, which meant that the imported lac had 'nearly as much colouring matter as one-tenth of its weight of cochineal'. But if the lac insects could somehow be separated from the resin 'the would afford as much colour and prove as valuable, as an equal weight of cochineal.' One of the weird ironies here is that the resin was a valuable commodity in its own right, providing the basis for shellac which, was increasingly used as a finish and veneer during the early nineteenth century. Bancroft, interested only in its colourant properties, was not disposed to see the gum as a potentially valuable by-product.

Ultimately the East India Company made the decision that its dyers could employ lac alongside cochineal—saving, Bancroft claimed, fourteen thousand pounds in the process [42],

without any inferiority in the scarlet so dyed, as far, at least, as I have been able to observe, or learn; and though, in a few instances, the cloths were injured by adhesions of the *resinous* part of these preparations, the injury was probably occasioned by their not having been so finely ground as they ought to have been.

More could be said about Bancroft's experiments—and, remember, this is only one chapter of his book, but I hope the general point is clear. Chemical techniques, 'knowledge of mordants' gave techniques for extracting colourant matter from the lac beetles but they could not overcome other complications in the production of this dye—its combination with resin and wax, the questions about availability of techniques and materials, and the powerful whims of the East India Company itself. Through all this, the guiding spirit for Bancroft's experiments was one of substitution: could the lac's colouring matter be made equivalent in brightness and value to that of the cochineal? This required investigation through multiple means, not simply control through chemical processes.

5.5 Conclusion

This paper has pursued the suggestion that eighteenth century chemistry should be regarded as a science of materials. I have tried to add to the story there in three ways. First, by looking not at formal taxonomic schemes but at the more messy accounts of materials as bundles of properties (some desirable, some not) which produced both uncertainties and productive opportunities in chemical practice. This included purification tests and forms of imitative invention. Second, by suggesting that *substitution*—as a way of trying to locate useful properties in new places—was an important motivator in this period across a range of different sciences, but that those sciences had slightly different practices around what were acceptable substitutive identities. Third, by arguing that chemists drew upon the unruly descriptive practices of certain kinds of natural history, an approach which I exemplified with the close reading of Bancroft's work. This leads me to two conclusions: In terms of *our* theory choice, it's helpful to think about the bundles of materials—wanted and loathed—with which chemists had to deal, and to see different scientific practices as ways to approach questions of desirable properties. This is why it is interesting to talk about a practice like substitution which employed different disciplines, like chemistry and natural history. That approach to materials complements and complicates the one which focuses on shifting taxonomic practices. Second, some of the theoretical choices of eighteenth century chemists significantly shaped by the effort to demonstrate identities or organize differences of useful commodities—this meant that they drew on a wider range of conceptual, institutional, and material resources than we might expect.

References

1. Klein U, Lefevre W (2007) Materials in eighteenth century science. A historical ontology. MIT Press, Cambridge, p 1
2. Roberts L (1995) The death of the sensuous chemist: the 'new' chemistry and the transformation of sensuous technology. Stud Hist Philos Sci Part A 26(4):503–529
3. Klein U, Lefevre W (2007) Materials in eighteenth century science. A historical ontology. MIT Press, Cambridge, p 296
4. Klein U, Lefevre W (2007) Materials in eighteenth century science. A historical ontology. MIT Press, Cambridge, pp 297–299
5. Spary EC (2000) Utopia's Garden. French Natural History from Old Regime to Revolution. The University of Chicago Press, Chicago
6. Brockway LH (2002) Science and colonial expansion. Yale University Press, New Haven
7. Berg M (2002) From imitation to invention: creating commodities in eighteenth-century Britain. Econ Hist Rev 55:1–30
8. Berg M (2005) Luxury and Pleasure in Eighteenth-Century England. Oxford University Press, Oxford
9. Delbourgo J (2009) Fugitive Colours: Shamans' Knowledge, Chemical Empire, and Atlantic Revolutions. In: Schaffer S, Roberts L, Raj K, James Delbourgo J (eds) The Brokered World. Go-betweens and Global Intelligence, 1770–1820. Science History Publications/USA, Sagamore Beach, chapter 7
10. Watt J (1784) On a new method of preparing a test liquor to shew the presence of acids and alkalies in chemical mixtures. Philos Trans R Soc Londo 4:419–422
11. Pugh JS, Hudson J (1985) The Chemical Work of James Watt, F.R.S. Notes Rec R Soc Lond 40:41–52
12. Chang H (2004) Inventing temperature: measurement and scientific progress. Oxford University Press, Oxford

13. Touwaide A (2011) Quid pro Quo: Reivisiting the practice of substitution in ancient pharmacy. In: Van Arsdall A, Timothy Graham T (eds) Herbs and Healers from the Ancient Mediterranean through the Medieval West: Essays in Honor of John M. Riddle. Ashgate, Aldershot, chapter 2

14. Dossie R (1758) The laboratory laid open; or the secrets of modern chemistry and pharmacy revealed. J. Nourse, London

15. Klein U, Lefevre W (2007) Materials in eighteenth century science. A historical ontology. MIT Press, Cambridge, p 110

16. Klein U, Lefevre W (2007) Materials in eighteenth century science. A historical ontology. MIT Press, Cambridge, p 111

17. Home F (1756) Essays on Bleaching. Dublin

18. Willich AFM (1802) The domestic encyclopaedia: or, A dictionary of facts, and useful knowledge, comprehending a concise view of the latest discoveries, inventions, and improvements, chiefly applicable to rural and domestic enconomy .... Murray and Highley, London, p 125

19. Brunello F (1973) The art of dyeing in the history of mankind. Vicenza, Neri Pozza, p 259

20. Mitchell J (1748) An account of the preparation and uses of the various kinds of Pot-Ash. Philos Trans (1683–1775) 45: 541–563

21. Stewart L (2008) The laboratory, the workshop, and the theatre of experiment. In: Bensaude-Vincent B, Blondel C (eds) Science and spectacle in the European Enlightenment. Ashgate, Aldershot, p 20

22. Page FC (2001) The Birth of Titrimetry: William Lewis and the Analysis of American Potashes. Bull Hist Chem 26(1):66–72

23. Dossie R (1768) Memoirs of agriculture: and other oeconomical arts. J. Norse, Dublin

24. Simon HA, Langley PW, Bradshaw GL (1981) Scientific discovery as problem solving. Synthese 47:1–27

25. Pickstone JV (2001) Ways of knowing: a new history of science, technology, and medicine. The University of Chicago Press, Chicago, chapter 3

26. Keyser BW (1990) Between science and craft: the case of Berthollet and Dyeing. Ann Sci 47(3):217

27. Evans C (2007) Crucible steel as an enlightened material. Paper presented at Steel in Britain in the Age of Enlightenment, University of Glamorgan, 7/8 December, p 7

28. Foucault M (1966) Les Mots et les choses. Editions Gallimard, Paris

29. Jardine N, Secord JA, Spary EC (1996) Cultures of natural history. Cambridge University Press, Cambridge

30. Spary EC (2005) Of nutmegs and botanists: the colonial cultivation of botanical identity. In: Schiebinger L, Swan C (eds) Colonial botany: science, commerce, and politics in the early modern world. University of Pennsylvania Press, Philadelphia, PA

31. Chenciner R (2000) Madder red: a history of luxury and trade. Routledge, New York

32. Delbourgo J (2009) Fugitive Colours: Shamans' Knowledge, Chemical Empire, and Atlantic Revolutions", In: Schaffer S, Roberts L, Raj K, James Delbourgo J (eds) The Brokered World. Go-betweens and Global Intelligence, 1770–1820. Science History Publications/USA, Sagamore Beach, pp 315, 318. For Bancroft's defence of a theoretical chemistry, see pp 312–314

33. Delbourgo J (2009) Fugitive Colours: Shamans' Knowledge, Chemical Empire, and Atlantic Revolutions", In: Schaffer S, Roberts L, Raj K, James Delbourgo J (eds) The Brokered World. Go-betweens and Global Intelligence, 1770–1820. Science History Publications/USA, Sagamore Beach, p 319

34. Bancroft E (1794) Experimental researches concerning the philosophy of permanent colours, and the best means of producing them, by dyeing, calico printing, &c. T. Cadell & W. Davies, London, p 36

35. Bancroft E (1794) Experimental researches concerning the philosophy of permanent colours, and the best means of producing them, by dyeing, calico printing, &c. T. Cadell & W. Davies, London, p 11

36. Bancroft E (1794) Experimental researches concerning the philosophy of permanent colours, and the best means of producing them, by dyeing, calico printing, &c. T. Cadell & W. Davies, London, p 12

37. Bancroft E (1794) Experimental researches concerning the philosophy of permanent colours, and the best means of producing them, by dyeing, calico printing, &c. T. Cadell & W. Davies, London, p 19

38. Bancroft E (1794) Experimental researches concerning the philosophy of permanent colours, and the best means of producing them, by dyeing, calico printing, &c. T. Cadell & W. Davies, London, p 27

39. Bancroft E (1794) Experimental researches concerning the philosophy of permanent colours, and the best means of producing them, by dyeing, calico printing, &c. T. Cadell & W. Davies, London, p 14

40. Bancroft E (1794) Experimental researches concerning the philosophy of permanent colours, and the best means of producing them, by dyeing, calico printing, &c. T. Cadell & W. Davies, London, p 15

41. Bancroft E (1794) Experimental researches concerning the philosophy of permanent colours, and the best means of producing them, by dyeing, calico printing, &c. T. Cadell & W. Davies, London, p 16

42. Bancroft E (1794) Experimental researches concerning the philosophy of permanent colours, and the best means of producing them, by dyeing, calico printing, &c. T. Cadell & W. Davies, London, p 36

Chapter 6
Theory Choice in Chemistry: Attitudes to Computer Modelling in Chemistry

Kat F. Austen

6.1 Introduction

Computational modelling has grown over the last 40 years into a widely applied methodology in the scientific community. The uptake in the chemistry, as in other sciences, has not always been smooth, although the movement gains momentum as increasing computer power correspondingly increases the complexity of computer simulations of chemical systems. Although as a subject chemistry is diverse and varied, computational chemistry simulations share one thing—they are rooted in the real world and necessarily non-ideal. This chapter details the historical progress of computational chemistry, the assumptions made in its most popular models, and some of the possible sources for objection within the chemical community.

6.2 Models in Chemistry

The distinction between model and theory in science has long been disputed. Traditionally, a model in chemistry is taken to be an image of a chemical system, be it the shell model of an atomic structure or a ball and stick model of a compound or crystal structure. These models can be realised, made tangible in order to add an experiential component to the understanding of the theory upon which they are based. By contrast a theory is a set of hypotheses about the world, formulated into a coherent explanation about how things work, which furthermore has some predictive component which can be tested against.

K.F. Austen (✉)
Artist in Residence, Faculty of Mathematical and Physical Sciences,
University College London, Gower Street, London WC1E 6BT, UK
e-mail: k.austen@ucl.ac.uk

© Springer International Publishing Switzerland 2016
E. Tobin and C. Ambrosio (eds.), *Theory Choice in the History
of Chemical Practices*, SpringerBriefs in History of Chemistry,
DOI 10.1007/978-3-319-29893-1_6

In recent decades, a phenomenon has increasingly pervaded chemistry—that of computational modelling. This is the use of computers to simulate chemical systems, and to predict behaviour—such that these models can be considered as a form of experiment. These models are based on the application of chemical theories to physical systems and make use of various types of approximations in order to generate a coherent model of the system. Broadly, there are 3 levels of abstraction of the computer model: interatomic potentials (force field methods), which models the interaction between atoms by use of mathematical functions that approximate bonding and non-bonding interactions; density functional theory, which models the probability distribution for electrons around point charges representing the atomic nuclei; and quantum mechanical calculations, which solve, with approximations, the Schrödinger equation for the system.

There are also various types of simulation to make use of these abstractions, from static simulations to find the lowest energy atomic configuration for the system, and dynamical simulations to find the equilibrium state for the system, but for the purposes of this study these distinctions are less important when discussing the theory choice made by chemists of accepting or not computational modelling into their practice.

6.3 Brief History of Computational Modelling in Chemistry

Computational modelling of chemical structures began with force field methods, the foundations for which were laid with developments in vibrational spectroscopy in the 1930s [1]. It was not until the mid-1940s that there began to be apparent a coherent theory, when three groups came up with similar methods for describing molecular conformations and their interactions with respect to sterics.

Force field methods use mathematical functions—interatomic potentials—to describe the attraction of different atom types to each other and the strain exerted on the molecular configuration by the presence of other atoms. They have become more elaborate over time, progressing from mainly non-bonded van der Waals interactions to more complicated 4-body terms and shell models of atomic distortion. Crucially, however, force field methods do not allow for drastic changes in the electronic configuration of a system—i.e.: bond making or breaking—and only describe the system well when the configuration is near equilibrium.

Bond making and breaking can be possible in computational modelling if electronic structure methods are used. There are two broad types of model within this family [2]. Ab initio (from first principles) methods start by trying to calculate the ground state wavefunction for the system—that is, given an input geometry by the chemist, the computation attempts to find the probability of finding electrons across the system by solving the Schrödinger equation for the system. However, many approximations must be made to do this for most systems as is not possible to solve the Schrödinger equation analytically for any but a two-body system such as the hydrogen atom.

One method for finding a numerical solution to a many-body problem is the Hartree-Fock method [3], devised in the late 1920s by Hartree [4] and refined by Vladimir Fock (though it wasn't until later that the equations were refined and implemented in computational code). Hartree-Fock methods are computationally very expensive, and inherent in the equations are a number of approximations that introduce artefacts in the computational model. These must be borne in mind when interpreting the results. Not least of them is the Born-Oppenheimer approximation, which separates the nuclear and electronic wavefunctions.[1]

By contrast, Density Functional Theory (DFT) [6] methods do not attempt to solve for the ground state wavefunction directly, but rather to find a universal function for a system's electron density and then to calculate each individual electron's density distribution within that system-wide density [7]. DFT dates from the mid-1960s [8] and, as with the other models, contains many approximations[2]— as do the various different optimisation and simulation choices for each of these methods, mentioned in Sect. 6.2.

There are many parameters to be refined for any electronic structure simulation, but one of great importance is the choice of basis sets. Basis sets are a set of mathematical functions that are used to describe the location of molecular orbitals in a computational chemical calculation. Choice of more elaborate basis sets requires greater computational expense but can result in a more accurate answer. (This is not always the case, however, as a cancellation of errors in more approximate calculations can sometimes result in an answer closer to real life than a more refined calculation.)

Simulation methodologies are applicable to different systems, and are chosen depending on variables such as system size, likelihood or necessity of bond breaking, and what type of information the researcher wishes to obtain. Most importantly, however, "success depends on the ability of a model to consistently reproduce known (experimental) data" [9].

Thus, there are numerous choices to be made when applying computational modelling once a chemical problem has been identified: the choice of model, and thereafter the choice of computer code to employ the model, for there are many, and choice between the variables that can be fine-tuned within the model such as basis sets, interatomic potentials or optimisation algorithm. But there is also a meta-choice: whether to rely on computational models at all. This latter choice is the focus of this chapter.

---

[1]The Born-Oppenheimer approximation has some limitations in accurately modelling chemical systems, for example where the ground and excited state are energetically close, or where the nature of bonding transitions between ionic and covalent. For further discussion see [5].

[2]For further details on computer simulation methods and the approximations therein see [5].

6.4 Reception and Appropriation of Computational Models in Chemistry

Until the 1950s computational models were rarely deployed in most areas of science. Since then, computer power has mushroomed approximately in accordance to Moore's law, which states that the number of transistors on a processor, with a proportionate increase in processing power, will double every 2 years. This has facilitated a blossoming of the field of computer simulation—the process of asking questions of a computational model of a chemical system—across many of the sciences, which began with nuclear physics.

Yet despite its growth computer simulation has been an area of controversy in many disciplines, championed by its users and maligned by those who fear a threat to traditional theory and experimentation—both on ideological and financial grounds.

Initial uptake of simulation methods was slow in many disciplines. In chemistry, it was the 1970s that saw a rise in simulations with a ubiquity of computer power, followed in the next decade by a real boom with the advent of personal computers and the emergence of graphical user interfaces which allowed chemists to actually visualise on the computer the chemical system that they were modelling [1]. The rise of High Performance Computing has meant that computational simulations can tackle increasingly realistic systems, even using ab initio dynamical techniques, to increase the scope of the field [10].

A good marker for acceptance of the techniques is their uptake in the commercial world. Fields like drug design and biochemistry, where modelling of isolated molecules is of use, were early adopters of the technique. In 1978, for example, the first commercial companies founded to use computational modelling for chemical purposes focussed on molecular design [11] and predictions of toxicity. Of course, a lack of commercial uptake cannot be taken to imply resistance to the methods, as there are many factors in the commercial viability of a particular approach. However, without confidence in the methods, it is unlikely that continued investment would be made in a venture. Simulations are now commonplace in these fields, with funding feeding into the academic community from pharmaceutical companies like Unilever and AstraZeneca.

There is also commercial confidence in the use of computational chemistry models in materials chemistry—modelling is one of the R&D core competences of Johnson Matthey [12], for example, a specialty chemicals company that has recently been enjoying considerable commercial success with a +20 % increase in revenue for the year 2011–2012 [13].

There was some latency in the uptake in the academic community of computational modelling, which was partly due to "unfamiliarity of the average chemist with the programming and use of mainframe computers" [14]; indeed the nuts and bolts of the theory and implementation were conspicuous by their absence. For example, nearly 20 years after molecular modelling methods really hit the map, introducing a discourse on the application of modelling to biomolecules in the journal Biochemical Education, by Christopher J Cramer, author of one of the most definitive teaching texts on computational methods, Essentials of Computational Chemistry, the editor CA Smith notes [15]:

> Despite the widespread use of computer-based molecular modelling, explanations of its essential background is conspicuous by its absence in the large, otherwise incredibly comprehensive, books which almost define Biochemistry.

Around the same time, computational chemist Delano P. Chong expressed hope—albeit possibly over-optimistically—that computational modelling techniques may become as ubiquitous as spectroscopic techniques, seeing them as a tool for the chemist [16]:

> Many experimental chemists use various kinds of spectroscopy in their research even though they are not spectroscopists. In a similar manner, more and more scientists are applying computational techniques as another weapon in their arsenal.

Not everyone was as hopeful about computational methods. A parallel can be drawn between the response to these methods in mathematics in the mid-1990s [16]. The community was split at the time by the introduction of computational methods to attempt to solve mathematical problems. The resistance arose mainly because of the opacity of the computer's process. As John Horgan writes [17]:

> Pierre Deligne of the Institute for Advanced Study, an algebraic geometer and 1978 Fields Medalist. 'In a way, I am very egocentric. I believe in a proof if I understand it, if its clear.' While recognizing that humans can make mistakes, he adds: 'A computer will also make mistakes, but they are much more difficult to find.'

While the mathematical community is far more wedded to the pursuit of pure theory than the chemical community, they share this reason for their resistance to computational methods—that the approximations and hard-to-detect errors of computer modelling erodes the researchers' faith in the methods. And there is some similarity in the two disciplines' recourse to computational methods—when addressing a system or problem that is analytically unsolvable.

6.5 Validation and Extrapolation

Computational models in chemistry were born of theory and are constantly refined according to empirical results. Clearly this is fertile grounds for philosophers of science—but it was not until recently that philosophers have begun to take notice of computational models [18]. However, the philosophical arguments over models and their application to chemical systems could hold the key to the reluctance with which some experimentalists and theoreticians have received the methods.

In searching for a coherent epistemology of simulations, philosophers are wont to refer to and draw analogies between theories and models, and models and experiment. The analogies can be helpful, which has prompted some philosophers to argue that there are no new questions raised by simulations [19], but such an approach can also be troublesome, as we shall see.

Computer simulations can be considered to be part of the scientist's search for truth [20]. But while model systems are ostensibly truth worthy, in as much as mathematics is a subject capable of true statements, in practice caution should be taken in suggesting that once implemented in code that is still the case. In coding up any mathematical model it is necessary to make approximations, whether this be the use of floating point numbers, or discarded terms in equations or approximation of functions.

Using the search for truth as a foil to determine the validity of a simulation approach can be helpful in understanding the often only half formulated exceptions that researchers may take to computer modelling: the applicability of the model and its reliability in generating accurate results for the real system it describes.

Under the surface of these exceptions lie the knotty problems of validation and verification. An important issue in all scientific endeavours, it is particularly problematic in simulation, where verification refers to determining whether the mathematical principles have been correctly encoded, and validation questions whether those principles are applicable to the system. As Boisvert et al. wrote when analysing scientific computation [21]:

> "Scientific software is particularly difficult to analyse. One of the main reasons for this is that it is inevitably infused with uncertainty from a wide variety of sources. Much of this uncertainty is the result of approximations. These approximations are made in the context of each of the physical world, the mathematical world, and the computer world." Where:
> "...neither the mathematical nor the computational model can be expected to be valid in all regions of their own parameter space."

Simulations are "downwards, motley and autonomous" [18]—they are in some respects a type of experimentation [22], in part drawn from theory[3] but also from other sources, and cannot be checked directly against observation—either because simulations are most often applied to systems we cannot measure experimentally, or because the simulations are abstractions which lack the complexity of a real-life system. With respect to epistemology, a computational model does not so much test chemical theory, but rather tests the predictions or hypotheses resultant from those theories, within the limitations of the approximations of the implementation in code.

These caveats make it troublesome to ensure that the results of a computer simulation are chemically relevant—when a mathematical system is so complex that it cannot be solved by a researcher, how is that researcher to validate the implementation of the theory? And when it cannot be directly compared to observation, how can the results be validated? Schmid argues that in the quest for truth, a simulation that is untestable against a real system—and thus is by default inaccurate—can still be "true" [19]. This, however, is likely to provide little solace to computational chemists. For pragmatists, the issue remains whether the results of simulations can answer questions about the real world.

---

[3]Experiments draw from and feed back into theory. Simulations differ in this regard, as the theory into which they feed back is that of simulation rather than theory of chemistry.

Should it be that the model can be tested against observable data things are nevertheless complicated [18]—if the model fails, how does the researcher ascertain if the problem is with the model, with its implementation into code, or with one of the many variables upon which the simulation relies? If it succeeds, how is the researcher to determine if it is indeed an accurate model and implementation or just a cancellation of errors?[4]

Parker argues that drawing from the philosophy of experiment can help define the epistemology of simulation such that interrogation of results from computer models should not only explicitly determine the canonical errors that are artefacts of the method, but they should actively investigate them [23]. Indeed, published results of computer simulations always detail the base-line methodologies, information that to a well-versed reader will outline the assumptions made. Furthermore, many papers make reference, in the discussion of the results, to the artefacts that might arise from the chosen methodology (e.g.: [24]).

Indeed, over time the methodologies for computer simulation gather their own credentials through repeated use, in much the same way that experimental procedures or apparatus do [18, 25]. Winsberg has provided a comprehensive treatise on philosophical comparisons between experiments and models [26], in which he argues that the it is the background knowledge behind either experiment or simulation that dictates whether either is reliable. The practitioner's background knowledge is also of interest here. Setting up a good computational simulation is often referred to as a "black art" among those in the field—a consequence perhaps of the seemingly inexhaustible variables involved in the process.

There are good and bad experiments, and the same is true of simulations. However, even though they share similarities, simulations are not experiment. Simulations only bear a theoretical relation between the model and the real-world system it describes, whereas the connection between real-world experiments and real-world systems is one of material abstraction, not virtuality. That extra level of abstraction brings with it a greater propensity for error—weaknesses in the joints between theory, hypothesis, experimentation and real-world phenomena—which are much harder to spot.

However the chemists' issues with simulation may lie deeper than the knotty arguments over the reliability of the models. Much simulation hinges on the acceptance of quantum mechanics as a means to describe chemical behaviour, effectively relying on reduction of chemical phenomena to physical laws—a topic of some discussion over the last few years and addressed adeptly by Robin Findlay Hendry recently [27]. And as described previously, computational models rely on

---

[4]This is particularly problematic: given that if a simulated system fortuitously accords with data from the real world given certain parameters, it could be from a cancellation of errors, chemists using such a simulation to predict how a system would behave if a parameter were changed have to trust that the model will also recreate these variations in the system accurately. There is often no way to check.

various approximations and work-arounds—necessary to reduce computational expense and to overcome the difficulties in solving a many-body Schrödinger equation. Indeed, applying quantum mechanics to real-world systems is so complex that quantum chemistry itself could not exist without computational methods.

And herein lies a problem. As theoretical physicist Paul Dirac famously stated in 1929 [28]:

> The underlying physical laws necessary for the mathematical theory of a large part of physics and the whole of chemistry are thus completely known, and the difficulty is only that the exact application of these laws leads to equations much too complicated to be soluble.

Tunnelling down, Hendry deftly summarises one of the main objections to the computational methods upon which quantum chemistry relies, held by generations of scientists [27]:

> ...quantum chemistry does not meet the strict demands of classical intertheoretic reduction, because its explanatory models bear only a loose relationship to exact atomic and molecular Schrödinger equations

Tunnelling up, on the other hand, there is the objection that chemistry as a subject is overall too complex for reductionism; where quantum chemistry cannot predict accurately real world phenomena—take, for example, the chemical similarity of vanadium and niobium which have different electronic structures [27, 29]. The same argument can be made against simulations that rely on interatomic potentials rather than electronic structure calculations; in either case the simulation will be too abstracted, too reduced from reality, to be useful.

While reductionism in science may be frequently debated in the halls of HPS departments, the fact remains that faith in the validity of reductionist approach pervades, indeed is not questioned within, the scientific research community at large. Two interesting elements arise when considering rejection of computational simulations of their subject matter by chemists, then. As discussed previously, the rejection may arise because computational methods cannot simulate something so complex as a real chemical system. This can be seen as a rejection of reductionism: we cannot understand the whole system by understanding the basic physical laws underlying it when the system is so complex. On the other hand, in the number of approximations that computational models must make in order to be able to model even a helium atom, they defy absolute reductionism. In this case, their rejection is in favour of reductionism.

Which argument rings true for which chemists is an open question. But with computational chemistry seeming to fail either way surely the only recourse is to pragmatism, and the quest for truth.

6.6 Conclusion

Computational modelling in chemistry is in a unique position compared to the other sciences. Where mathematics and physics are happier to study ideal systems, real-life chemical systems are more complex and often non-ideal, in a similar way to systems at the biological scale and greater. However, the scale of chemical interrogation of these systems often necessitates inclusion of electronic and quantum effects, which increase the computational cost of the simulation and the number of approximations necessary.

Much of the distrust of computer simulation arises from reservations about the applicability of these approaches to chemical systems, and the validity of the results obtained by them. Nevertheless, as computer power increases and models become more complex and realistic some of these reservations are addressed and computational chemistry is increasingly accepted.

Some fundamental issues inherent in computational chemistry remain, such that they cannot be conceived of to be just another experimental tool for the chemist. However, in practice the models serve a real and useful purpose, and the main challenge is in encouraging the chemical community to engage in the philosophical discourse and bear it in mind when addressing their work.

References

1. Schelct MF (1997) Historical Overview of Molecular Modelling. John Wiley & Sons, Hoboken
2. Catlow CRA (2005) Computer modelling in materials chemistry. Pure Appl Chem 77:1345–1348
3. Cramer CJ (2002) Essentials of Computational Chemistry. John Wiley & Sons, Hoboken
4. Hartree DJ (1928) The wave mechanics of an atom with a non-coulomb central field. Part I. Theory and methods. Math Proc Cambridge Philos Soc 24(1):89–110
5. Jensen F (1999) Introduction to Computational Chemistry. John Wiley & Sons, Hoboken
6. Parr RG (1983) Density functional theory. Annu Rev Phys Chem 34:631–656
7. Gross EKU, Dreizler RM (1995) Density Functional Theory. Plenum Press, New York
8. Kohn W, Sham LJ (1965) Self-consistent equations including exchange and correlation effects. Phys Rev 140:A1133–A1138
9. Hehre WJ (2003) A guide to molecular mechanics and quantum chemical calculations. Wavefunction, Inc, Irvine CA
10. Woodley SM, Catlow CRA (2011) High-performance computing in the chemistry and physics of materials. Proc Roy Soc A 467:1880–1884
11. Richon AB (2001) A scrolling history of computational chemistry. https://www.researchgate.net/publication/280934406 (Rretrieved 10 March 2016)
12. Johnson Matthey, Research and development (2012a) http://www.matthey.com/innovation/innovation_in_action/accurate-modelling-advanced-products (Retrieved 30 December 2012)
13. Johnson Matthey, Research and development (2012b) http://www.matthey.com/investor/reports (Retrieved 10 March 2016)
14. Schelct MF (1997) Historical Overview of Molecular Modelling. John Wiley & Sons, Hoboken

15. Smith CA (1994) Problem-based learning. Biochem Educ 22(3):14
16. Chong DP (1995) Recent Advances in Density Functional Methods Part 1. World Scientific, London
17. Horgan J (1993) The death of proof. Sci Am 269(4):92–103
18. Winsberg E (2009) Computer simulation and the philosophy of science. Philos. Compass 4 (5):835–845
19. Frigg R, Reiss J (2009) The philosophy of simulation: hot new issues or same old stew? Synthese 169:593
20. Schmid A (2005) What is the truth of simulation. J Artif Soc Soc Simul 8(4):5
21. Boisvert RF, Cools R, Einarsson B (2005) Assessment of accuracy and reliability in accuracy and reliability in scientific computing. Soc Ind Appl Math
22. Morgan MS, Morrison M (1999) Models as Mediators: Perspectives on Natural and Social Sciences. Cambridge University Press, Cambridge
23. Parker W (2008) Computer simulation through an error-statistical lens. Synthese 163(3):371–384
24. Austen KF, White TOH, Marmier A, Parker SC, Artacho E, Dove MT (2008) Electrostatic versus polarisation effects in the adsorption of aromatic molecules of varied polarity on an insulating hydrophobic surface. J Phys: Condens Matter 20:035215
25. Winsberg E (2003) Simulated experiments: methodology for a virtual world. Philos Sci 70:105–125
26. Winsberg E (2010) Science in the Age of Computer Simulation. University of Chicago Press, Chicago
27. Hendry RF (2012) Reduction, emergence and physicalism. In: Woody AI, Hendry RF, Needham P (eds) Philosophy of Chemistry. Handbook of the Philosophy of Science, vol 6. North Holland, Amsterdam, pp 367–386
28. Dirac PAM (1929) The quantum mechanics of many-electron systems. Proc Roy Soc London A123:714–733
29. Scerri E (1997) Has the periodic table been successfully axiomatised? Erkenntnis 47:229–243

Printed in the United States
By Bookmasters